Generis
PUBLISHING

REINFORCEMENT OF A BRICK MASONRY BUILDING

Petraq Koka, Doctor of Technical Sciences

Title: REINFORCEMENT OF A BRICK MASONRY BUILDING

ISBN: 979-8-89966-124-2

Author: Petraq Koka, Doctor of Technical Sciences

Cover image: www.pixabay.com

Publisher: Generis Publishing
Online orders: www.generis-publishing.com
Contact email: info@generis-publishing.com

CONTENTS

INTRODUCTION

Our country has inherited a significant number of residential buildings featuring brick retaining walls, primarily constructed from the 1970s to the 1990s. These structures bear the limitations and challenges typical of that era, reflecting the broader context of our nation's exclusion from advanced technologies and its struggles with extreme poverty and scarcity of construction materials. Among the primary shortcomings of these buildings are:

1- The residential buildings from the 1970s to 1990s in our country exhibit significant shortcomings, primarily stemming from the inferior quality of construction materials used.
   a- Firstly, there is uncertainty regarding whether the bricks utilized adhered to established standards during production.
   b- Additionally, the mortar, often made from river sand with high dust content, failed to achieve the necessary strength.
   c- The concrete used in these constructions also lacked proper standards and post-curing treatment, compromising its durability.
   d- Lastly, the floor slabs were predominantly prefabricated with minimal reinforcement, contributing to the overall structural weaknesses of these buildings.

2- Exposure to natural elements has significantly affected these buildings. Many are either
unplastered or poorly plastered, making them vulnerable to weather-related damage that compromises the integrity of the brick walls.
3- Compliance with Eurocodes and modern building norms is absent, as these regulations were not in effect during the construction of these buildings. However, it is important to note that they adhered to the Technical Design Conditions (KTP) that were relevant at the time.
4- The foundations of these structures exhibit considerable deficiencies, primarily constructed with unprocessed stone walls, and often lack the requisite horizontal concrete belts at the recommended heights per KTP guidelines.
5- Numerous structural modifications have been implemented on these buildings, often without the guidance of engineering expertise, leading to significant deviations from the original design intent.

These concerns have been a persistent focus of mine, prompting an analysis of the stability of these structures to address two key questions:

1) How resilient are these buildings in the event of a design-level earthquake?
2) What are the costs associated with upgrading them to comply with safety standards as outlined in the Eurocode?

The goal of this study is to address the aforementioned questions.
It will begin with a comprehensive overview of the standards and specifications that bricks and mortar must adhere to, accompanied by tables detailing the relevant indicators. Subsequently, the foundational criteria for calculating and assessing the stability of these buildings will be presented. To effectively illustrate these principles, I will focus on a specific case study: Building 77/5, indexed as such by the Civil Engineering Institute of its time, which has been utilized repeatedly across various prefectures in Albania.

CHAPTER 1

NORMS FOR MASONRY CONSTRUCTIONS

1.0 DEFINITIONS
1.1 STRUCTURAL ORGANIZATION

Buildings featuring vertical load-bearing structures constructed from masonry systems, which are integrated with horizontal elements in both the floors and foundations, must adhere to specific technical requirements.
These buildings must be conceived as a three-dimensional structure. The resistant systems of the masonry walls, the horizontal elements and the foundations must be connected to each other, in order to resist vertical and horizontal actions. The organization of the entire structure must be such as to ensure a total "box" behavior. All walls must be connected at the level of the slabs by reinforced concrete floor belts. The floors of the building must be, as much as possible, compact and symmetrical in relation to the perpendicular axes.
Buildings should have as many structures as possible characterized by regularity in plan and height. If necessary, it can be continuously divided by joints (connections) into structural subdivisions, dynamically independent of each other.
- A building is regular in plan, if all the following conditions are respected:
a) The configuration in plan is compact and approximately symmetrical in the two orthogonal directions, with respect to the distribution of mass and rigidity.
b) The ratio of the edges of a rectangle in which it is inscribed is less than 3
c) No dimension of any recess in the floor plan should exceed 25% of the overall dimension of the building in the corresponding direction.
d) Horizontal elements can be considered infinitely rigid in their plan, with respect to vertical elements.
- A building is regular in height, if the following conditions are met:
a) All vertical resistant elements (such as frames and walls), extend to the entire height of the construction.
b) The mass and stiffness of the building should be uniform or transition gradually from the base to the top, ensuring structural integrity. Specifically, the variation in mass between horizontal planes must not exceed 25%, while stiffness should not decrease by more than 30% or increase by more than 10% from one plane to the next. This continuity is essential to maintain the overall stability and performance of the masonry systems under load.

1.2 VERTICAL MASONRY STRUCTURES

Two main types of vertical masonry structures are distinguished: ordinary and reinforced.

Ordinary masonry - It is a combination of artificial or natural elements placed in a regular manner and connected to each other with mortar.
They can be considered as single walls, if the walls are without cracks and connected in a continuous vertical manner, or double, if the monolithic and solid connection of the walls is guaranteed.
Vertical bearing systems (walls) must withstand vertical and horizontal loads. Walls perform a bearing function when they are mainly under the influence of vertical loads and have a bracing function when they are mainly under the influence of horizontal loads. For the most correct static and dynamic behavior of the building, all walls should have, as much as possible, both the load-bearing and the bracing function.
If the transverse wall has openings (doors or windows) it is conventionally accepted that the stiffening function can be performed when the openings (cracks) are surrounded by a full wall, with a length of at least **0.2-0.3** of the height of the wall itself.
Each wall must be connected to transverse walls, which also perform the function of limiting second-order phenomena. These are controlled through the conventional flexibility of the wall. (see Tab. WALL 7)
The geometry of earthquake-resistant walls, in order to avoid possible effects of local instability, due to its actions, must respect, in function of the construction technology, the requirements of minimum thickness and flexibility, by limiting the ratio between the height of free bending and the thickness of the wall, as well as between the maximum height of consecutive openings in the masonry and the length of the wall.
The thickness of the walls must not be less than the values of Tab. 3.2. Buildings with ordinary masonry, must, as a rule, have openings arranged vertically. Otherwise, only the parts of the wall that present vertical continuity from the floor under verification to the foundations will be taken into account (in the structural model and in the checks to be carried out). Appropriate models and analysis methods may be used to approximate the overall behavior of the walls.

1.3. **HORIZONTAL STRUCTURES**

They can be floor slabs made of reinforced concrete, prestressed or mixed structures. They must solve the function of distributing horizontal actions between the structural walls, above all, they must be well connected to the walls and guarantee a very good diaphragm function. The maximum distance between two consecutive slabs must not be exaggerated. The horizontal constituent structures and the coverings must not have lateral thrust effects. Any horizontal thrusts, evaluated taking into account seismic actions, must be absorbed by suitable structural elements. The horizontal elements are, according to the standard, floor slabs, or inclined coverings, which must ensure, from the point of view of resistance and rigidity, the distribution of horizontal actions on the bracing walls.

1.4. **FOUNDATIONS**

When natural elements are used, stones with almost parallelohedral geometry, placed in the work in regular layers, the masonry is called regular (squared). For seismic zones 6 and 7, the placement of unprocessed stone elements may be allowed, provided that combined masonry is realized, consisting of irregular stone masonry, interrupted by simple or reinforced concrete strips, or at least 2 complete rows of bricks placed horizontally, at an axial distance of no more than 1.6m and extending throughout the length and thickness of the wall. The foundation structures must be realized with reinforced concrete and verified based on the stresses provided by the analyses. They must be continuous, without interruption, in accordance **with the cracks of the supporting walls.**
If a basement or semi-basement floor with reinforced concrete walls is present, it can be considered as a foundation structure for the load-bearing masonry floors above it, if it meets the requirements for foundation continuity.
The use of unprocessed stone materials is permitted only in seismic zones 4 and 5, provided that they are placed in the work in approximately regular layers.

1.5. **CONCRETE BELTS**

The walls are generally completed at their top with horizontal reinforced concrete elements as wide as their width (girders). The height of the girders must be the same as that of the floor slab, but not less than 15 cm.
- The reinforcement must be no less than 8cm², with a diameter of no less than 16 mm;
- The rods must have a diameter of no less than 6mm and an interaxle of no more than 25cm.

The connection between the foundations and the structure is also made through b/a girders, placed at the base of each vertical bearing masonry, with a width and height of no less than 1/2 of the width of the respective thickness.
If the mezzanine consists of prefabricated slabs and beams, as well as metal beams, they must be supported on the girder, for a width of no less than half its width, but not less than 12 cm; Metal beams must be equipped with appropriate anchorages.

1.6 REINFORCED MASONRY

Reinforced masonry is considered to be masonry constructed with semi-solid artificial elements, with vertical holes, (coaxial) which allow the placement of vertical reinforcement. Horizontal reinforcement may be distributed on the surface of the mortar between the masonry elements. The mortar, or the conglomerate filling the holes, or that of the layer where the reinforcement is placed, must have Rck ≥ 15 N/mm2 and must cover all the reinforcement. The thickness of the covering must be such as to guarantee the transmission of stresses between the masonry and the reinforcement and to form an adequate covering of the reinforcement. In the case of buildings with reinforced masonry, each wall constitutes, in its complexity, a structure with holes, from the openings in it. All the walls must be connected to the slabs in such a way as to constitute rigid diaphragms; the resulting structure must be able to respond to external horizontal actions with a general type of behavior, to which the resistance of the walls only contributes at the respective floor.

In the case of reinforced masonry elements, the vertical and horizontal reinforcement is dimensioned in analogy to what is foreseen for reinforced concrete elements. The reinforcement bars must be such as to ensure the required adhesion and embedment. Adequate protection of the reinforcement against corrosion must be guaranteed.

The horizontal reinforcement placed in the mortar layer, or in the particular cracks of the blocks, must be distributed according to the respective recommendations and in such quantities as to avoid both tensile cracking (minimum percentage) and destruction by compression of the masonry (maximum percentage).

Vertical reinforcement must be placed in specific gaps or cracks, with appropriate dimensions, to guarantee the transmission of stresses to the masonry elements. Vertical reinforcement, with appropriate cross-section, must be placed at each extremity of each load-bearing wall, at each joint (interruption) of the load-bearing walls, in correspondence of each crack, but nevertheless in a limited interaction, in order to absorb the occasional localized tensile and compressive stresses, and to guarantee a unified behavior of the wall.

CHAPTER 2- Materials and typological characteristics

2.1 MORTAR

Descriptions regarding masonry mortar are given below, **Tab 3.1.**
Masonry mortars must guarantee appropriate parameters in terms of durability and mechanical performance and must have certificates in accordance with European standards EN998-2. It is necessary that the components of the mixture do not contain organic substances, powdered oils or clays. The mechanical properties of a mortar are determined by its average compressive strength ***fm***.
The category of mortar is determined by a code starting with the letter M followed by a number that under understands the resistance *fm*, given in N/mm2 according to **Tab 3.1.** It is not allowed the use of mortars with:

$$\mathbf{fm \leq 1\ N/mm2} \tag{2.1}$$

2.2 RESISTANT ARTIFICIAL ELEMENTS IN MASONRY

For artificial elements (made of brick or concrete) to qualify as having resistant functions, they must meet the criteria outlined in **Tab. 3.9**
These elements may have cracks in the direction perpendicular to the support surface (elements with vertical holes). For use in works that comply with current standards, the elements are classified by the percentage of cracks **φ** and the average cross-sectional area of each crack ***f.*** The holes must be distributed almost uniformly over the surface of the element.
The percentage of cracks is given by the formula:

$$\mathbf{\varphi = 100F/A} \tag{2.2}$$

where:
F- Surface area of penetrating holes throughout the element and non-penetrating holes.
A- Total horizontal surface area of the masonry element, limited by its perimeter.

Tab. **Tab. 3.2** gives the classification of brick and concrete elements.

2.3 MECHANICAL CHARACTERISTICS OF MASONRY:

The basic properties on the basis of which masonry is classified are:
• characteristic compressive strength *fk*;
• characteristic shear strength, without the presence of axial force f_{vko};
• modulus of elasticity in vertical shear E;
• modulus of elasticity in tangential shear G.

The characteristic strengths f_k and f_{vko} are determined either experimentally, on masonry samples, or with some limitations in function of the qualities of the components.
For projects in which the control of the stability of the masonry requires a value **fk ≥ 8 N/mm2**, it is mandatory to control the value of **fk**.

2.4 - STRUCTURAL ANALYSIS

For each specific control limit state, a computational model must be adopted which must:

- provide an accurate description of the organization of the structure, materials and location of the building;
- describe the most realistic behavior of the entire structure;
- consider the actions foreseen by the project and their application method to be realistic.

For the assessment of local effects, the use of computational models is allowed.
The response of the model is calculated using:

- linear analysis, taking the cut-off values of the elastic moduli.
- nonlinear analysis;

The analysis must provide for each structural element the following data:

- the axial force as a result of vertical loads; for buildings with a height greater than **10m**, the change in the axial load from horizontal actions must be taken into account;
- the shear force arising from vertical and horizontal loads;
- the eccentricity of the axial loads;
- the bending moment arising from vertical and horizontal loads.

The use of simplified models, based on the scheme of complete joints (hinges), is allowed for the calculation of the loads transmitted from the slabs to the walls and for the evaluation in the latter of the effects of out-of-plane actions.
Under the hypothesis of infinite stiffness in the plane of the slabs, the model can consist only of continuous wall elements from the foundation to the top, which allow horizontal displacements **only in the plane of the slabs**.
The out-of-plane reaction of the walls is to be considered in general as a local behavior independent of the general reaction and derived from the reaction of the wall in its own plane. Therefore, out-of-plane controls should be made separate and for them the equivalent forces shown for non-structural elements can be adapted.

2.5 LOADS AND THEIR COMBINATIONS

The design loads are those provided for in the EC.
The loads in the structure must be grouped in such a way as to create the most unfavorable loading conditions at the end of each verification. The probability of

simultaneous action of all actions is reduced, with the respective most unfavorable values.

2.6 – CONTROLS

The checks must be made according to the serviceability limit state and the ultimate limit state.
The hypothesis is accepted that the plane sections remain plane, the tensile strength of the masonry is not taken into account. Each wall section must be verified according to the **ultimate limit state or stress state** for the following loading conditions:

a) in compression-flexure for lateral loads;
b) in compression-flexure for floor loads;
c) in shear for floor loads;
d) for concentrated loads

Verifications according to **serviceability limit states** can be avoided, assuming that:
• the need to limit the size of cracks, for the permanent combination under the hypotheses:
- tensile resistance is assumed to be zero
- stress distribution is linear
- cracks are limited within the thickness of the plaster;
• for buildings with a number of floors greater than 4, the displacement of the floors ***dr*** must be checked, which for the most common combinations must respect the condition:

$$\mathbf{d_r \leq 0.003\ h} \qquad (2.3)$$

where **h**-is the height of the floor.
The theoretical explanation of the checks is given below in **CHAPTER 4.**

2.7. DETERMINATION OF MASONRY MECHANICAL PARAMETERS

2.7.1 COMPRESSIVE RESISTANCE. Its experimental determination

The characteristic experimental compressive resistance is determined by tests with ***n*** walls ($n \geq 6$), (see **Tab. 3.4**). The determination of the characteristic resistance must also be completed with the verification of the materials:
- mortar, (see Tab. **3.1**);
- masonry components, (see **Tab. 3.2** and **TAB. 3.3**);

2.7.2 COMPRESSIVE STRENGTH ASSESSMENT

In the design center, for masonry constructed from full or semi-full artificial elements, the value of f_k can be derived from the compressive strength of the elements and from the mortar group, through **Tab. 3.9.**
For values not included in the table, the use of linear interpolation is allowed.

2.7.3 SHEAR STRENGTH. Its evaluation in the absence of normal stresses.

In the design center, for masonry composed of full or semi-full artificial elements, the value of f_{vko} can be derived from the compressive strength of the elements, according to Tab MUR.6/A.
The validity of this table is limited to those masonry that have horizontal and vertical joints filled with mortar, with a thickness of 5 to 15mm. For values not included in the table, the use of linear interpolation is allowed.

2.7.4 BEAMS IN MASONRY

The shear resistance Vt of the connecting beams, in the presence of a floor joist or a bending-resistant lintel, well anchored at the extremities, is calculated in a simplified manner with the formula:

$$\mathbf{V_t = h_t\, f_{vdo}} \qquad (2.4)$$

- **CHAPTER 3**
  DESIGN OF MASONRY IN THE PRESENCE OF SEISMIC ACTIONS

3.1. SAFETY CHECK

For the safety control in the ultimate limit state, the resistance of each structural element to earthquakes must result greater than the active action for each of the following collapse modes: bending, shear and sliding in the plane of the wall, out-of-plane bending.
All walls that have a structural function, especially those that carry vertical loads, when they are not considered resistant to seismicity, must be subject to out-of-plane compression-bending control. In the case of application of the principle of the hierarchy of resistance (reinforced masonry), shear control will be a derivative of the compressive-flexural resistance, so that such a damage mechanism precedes a more easily sheared fracture.

3.2. SPECIFIC CHARACTERISTICS of masonry buildings in seismic zones

3.2.1 MATERIAL REQUIREMENTS

The elements to be used for load-bearing masonry buildings must be such as to avoid breakage (immediate rupture). For this purpose, the elements must meet the requirements given in the previous paragraph, with the following limitations:
- The volumetric percentage of eventual voids shall not exceed 45% of the total volume of the block
- The elements shall be placed parallel to the wall plane, shall be continuous and placed in a straight line; the only interruptions (discontinuities) that are accepted are those that correspond to the holes for their capture, or for the placement of the reinforcement
- The characteristic resistance ($\mathbf{f_{bk}}$) to fracture, in the bearing direction, calculated on the gross surface, shall be not less than **5N/mm²,** (5MPa);

- The characteristic breaking strength ($\mathbf{f_{bk}}$), in the direction perpendicular to the bearing, i.e. in the development plan of the walls, calculated in the same way as above, shall be not less than **1.5N/mm²** (1.5MPa)
- The mortar for ordinary masonry must have an average resistance of not less than **5Mpa** and the vertical joints must be filled with mortar.
- The use of material, or types of walls that have characteristics other than those referred to above, must be authorized by the Central Technical Service (STQ), under the approval of the Superior Council of Public Works.
- It is advisable to use masonry with unprocessed stones, with brick strips, only in areas with low seismicity.

3.2.2. DISTANCE BETWEEN ADJACENT BUILDINGS

The distance between adjacent buildings must be such as to avoid the phenomenon of impact (collision), therefore it cannot be less than the sum of the maximum displacements determined by the calculations. In any case the distance between two opposing points cannot be less than 1/100 of the height of these points, measured from the foundation plan, multiplied by:

$$\mathbf{ag \cdot S / 0{,}5g \leq 1.}$$

When specific calculations are not carried out, the maximum displacement of a building, without (seismic) isolation on the foundation, can be estimated at 1/100 of the height of the building, multiplied by ag· S/0.5 g.

3.2.3 MAXIMUM HEIGHT OF NEW BUILDINGS

- For structural typologies: - buildings with unreinforced masonry, corresponding to the elastic reserves of the structures, falling into zone 1 (high seismicity), a

maximum height is fixed, equal to two floors from the ground level, or from the street level. The second floor roof slab, cannot be used for additional (volume) habitable.
- For other zones, the maximum height of buildings must be limited in function of their deformative and absorbing (absorbing) capacity, as well as the seismic classification of the zone.
- For other structural typologies (reinforced concrete, iron, etc.) the maximum height is determined only by the resistive and deformative capacity of the structure.

3.3 SIMPLIFIED VERIFICATION OF MASONRY BUILDINGS

For buildings made of masonry made of full or semi-full artificial resistant elements, or from natural resistant elements, which are built in seismic zones of category 6-7, when they can be called "simple" constructions, a simplified verification can be carried out, accepting the coefficient $\gamma_M = 4.2$
A masonry building, to be called simple, must respect the following conditions:

a) The main walls of the structure must be continuous from the foundation to the roof.
b) No floor height shall exceed 3.5 m
c) No. of floors shall be greater than 3 (inside and outside the ground) for buildings with ordinary masonry and 4 for buildings with reinforced masonry.
a) The thickness of the masonry, according to the expression $l = h_0/t$, shall in no case be greater than 12
b) The variable load for the slab shall not be greater than 3.00 kN/m^2
c) The area of the masonry section resistant to horizontal actions, expressed as a percentage, in relation to the total area of the building floor, shall not be less than **4%** in the two main directions excluding protruding parts; walls with a length L of less than 50 cm, measured without including openings, are not to be taken into consideration for calculating the percentage of resistant masonry.

3.3.1 LATERAL LIMITING FACTOR

h/a	**ρ**
$h/a \le 0.5$	1
$0.5 \le h/a \le 1.0$	$3/2 - h/a$
$1.0 \le h/a$	1/[1+(h/a)2]

Where $\mathbf{h_0 = \rho h}$ is the free bending length of the wall, estimated based on the conditions of the connections at the ends, expressed in table 2 and **t** is the thickness of the wall.
The following conditions must also be respected:
- In each of the two directions, at least two wall systems must be provided with

their total length, excluding openings, each not less than 50% of the building dimension in the direction in question.
- In the calculation of the total length, only walls that comply with the geometric requirements of Tab. 7.8.II may be included. The distance between these two wall systems (walls), in the orthogonal direction to their longitudinal development on the floor, must not be less than 75% of the building dimensions in this direction. At least 75% of the vertical loads shall be supported by walls that are part of the system resistant to horizontal actions.

- In each of the two directions, walls resistant to horizontal actions must be present, with accial distance not greater than 7 m, and a height of up to 9 m for reinforced masonry constructions
- For each floor, the ratio between the surface area of the resistant section of the walls over the gross surface area of the floor, as a function of the number of floors of the building and the seismicity of the area, must not be less than the values given in **Table 7.8 III**, for each of the two orthogonal directions.
It is implicitly accepted that, for simple constructions, the number of floors cannot be greater than 3, for ordinary masonry constructions and 4 for reinforced masonry constructions.

$$\mathbf{S=S_S*S_T} \tag{3.1}$$

S_S= 1.234, stratigraphic coefficient
S_T=1, topographic coefficient.
-S_T it is applied only in the case of structures of use class III and IV.
Above all, for each floor it must result in: $\sigma = \frac{N}{A} \leq 0{,}25\frac{f_k}{\gamma_M}$ where:
• **N** is the total vertical load on each floor of the building corresponding to the sum of permanent and temporary loads (estimated on the assumption γ G = γQ = 1)
• **A** is the total surface area of the bearing walls of the same floor
• **f_k** is the characteristic compressive strength in the vertical direction of the masonry
The dimensioning of the foundations can be done in a simplified way, taking into account the average normal stresses and the total seismic oscillations determined with linear static analysis.
The dimensioning of foundations can be done in a simplified manner, taking into account the average normal stresses and total seismic oscillations determined with linear static analysis.

3.4. EXAMPLE OF SIMPLIFIED SEISMIC VERIFICATION of a masonry building

Let us analyze a simple building, with solid brick masonry, as in the figure, which has **f_k** = 7.5 N/mm^2, located in a flat area, on category B soil, to perform a simplified verification.

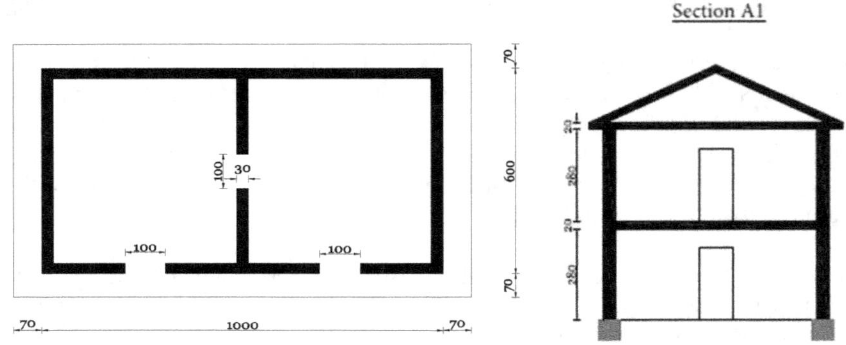

Fig. 3.1 – Example building analysis

1- Let **first verify** the existence of the simplicity conditions required to enable the continuation of the simplified control.

Points **a-c** are satisfied (the height of the building is 2 floors):

Verify point **d**:

• The ratio between the width and length of the object:

$\frac{length}{width} = \frac{10}{6} = 1.67 < 3$; point **d** is completed

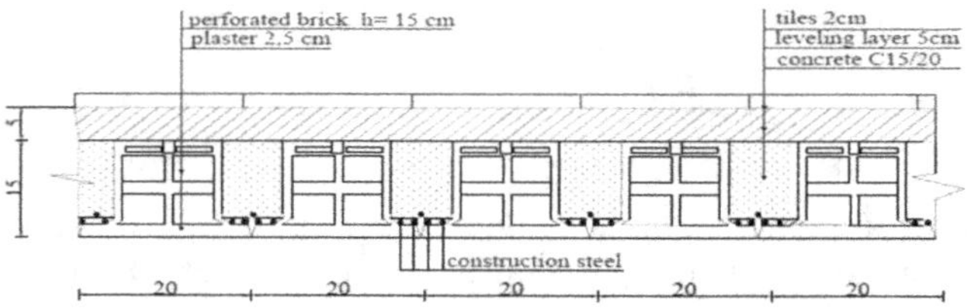

Fig. 3.2 – Slab section

2-Verify the point **e**

Flexibility is calculated according to the formula $\lambda = \frac{h_0}{t}$

Where: t = wall thickness

h_0 = free length of lateral wall bending, calculated through $h_0 = \rho \times h$

h= internal floor height

ρ = lateral connection factor

We must verify that none of the walls has a flexibility greater than 12, for the ρ factor we accept the value 1, which is preventive and valid for insulated walls.

In the case under analysis, we will have:

$$h_0 = \rho \cdot h = 1 \cdot 280 = 280cm$$

Flexibility will be: $\lambda = \frac{h_0}{t} = \frac{280}{30} = 9{,}33 < 12$;

Even the "e" point is completed.

We verify point g:

The surface of the resistant masonry in the direction X:

2 x 0.30 x 10.00 = 6,00

Remove: 2 x 0.30 x 1.00 = 0,60

Total = 5,40 m^2 ;

$$\frac{Resistant\ surface}{Building\ surface} = \frac{5.40}{6.00 \cdot 10.00} = 0.09 = 9\% > 4\%$$

The surface of the resistant masonry in the direction Y:

3 x 0.30 x 6.00 = 5,40

Remove : 0.30 x 1.00 = 0,30

Total = 5,10 m^2

$$\frac{Resistant\ surface}{Building\ surface} = \frac{5.10}{6.00 \cdot 10.00} = 0.08 = 8\% > 4\%$$

The minimum ratio between the bearing walls and the building surface, in both directions, is greater than the required one, **4%**. Point **g** is verified.
We now verify that the resistant surface of the walls, in each direction of verification, is greater than that predicted by table 7.8;
For this purpose, we proceed to calculate ag :

- ag = **0.317 g**
- stratigraphic coefficient Ss = 1.234,
- topographic coefficient ST=1.

And we have :

$$S*a_g = S_S*S_T*a_g = 1{,}234*1*0{,}317\ g = 0{,}39\ g;$$

Thus from table 7.8 we note that the percentage of masonry must be greater than 6.5%.
Previously we found that the percentages of masonry are 9% in direction X and 8% in direction Y, i.e. the conditions are met.
Assuming all other conditions are verified, we proceed with the simplified seismic verification.

• LOAD ANALYSIS

Since the coefficients γG and γQ must be equal to 1, no distinction is made between permanent and temporary loads. Let calculate the loads per m^2.

Structural Unit	LOAD NAME	WEIGHT daN/m2
ROOF	Roof self-weight	150

	Temporary load	100
	Amount	**250**
SLAB II	Trav weight: 2 x0.10x0.16x2500	80
	Slab 0.05x2500	125
	Bricks	80
	Plaster	30
	Temporary load:	100
	Amount	**415**
SLAB I	Trav weight: 2 x 0.10 x 0.16 x 2500	80
	Slab 0.05 x 2500	125
	Bricks	80
	Plaster	30
	Slab 0.05x2000	100
	Floor	40
	Light partition (brick)	100
	Temporary load:	200
	Amount	**755**

• Determination of the normal stress "N" on the foundation base

Calculation of the total weight,

Roof + Slab II 7.40 x 11.40 x (250 + 415) = 56.09 daN

Weight of masonry First Floor 3 x 0.30 x 2.8 x 6.00 x 1.800 = 27.216 daN

2 x 0.30 x 2.80 x (10.00 - 3 x 0.30)x1800 = 27.518 daN

Weight of Slab I 6.00 x 10.00 x 755 = 45.300 daN

Masonry weight, Ground floor: 3 x 0.30 x 2.80 x 6.00 x 1.800 = 27.216 daN

3 x 0.30 x 2.80 x (10.00 - 3 x 0.30)x1800 = 27.518 daN

Total = **211.75daN**

We calculate: $[\sigma] = \frac{N}{A} \leq \mathbf{0,25}\frac{f_k}{\gamma_M} = \mathbf{0,25}\frac{\mathbf{7,4}}{\mathbf{4,2}} = \mathbf{0,44} \quad N/mm^2$

Calculate the total area:

Masonry area in the direction **X:**

2 x 0.30 x 10,00 = 6,00

2 x 0.30 x 1.00 = - 0,60
Total = 5,40

Masonry area in the direction **Y**:

3x 0.30 x (6.00-2 x 0.30) = 4,86
-0.30 x 1.00 = - 0,30
Totali = 4,56
Total Area = 9.96m²

$$\sigma_m = \frac{N}{A} = \frac{211.752}{9.96} = 21.260\frac{dN}{m2} = \frac{0,21N}{mm^2} < 0,44\ N/mm^2$$

Verification completed.

TAB. 3.2 (WALLS. 2/A) - TYPOLOGY OF ELEMENTS AND MINIMUM WALLS THICKNESSES

Element type	φ	Brick		Beton	
		f (cm²)	minimum thickness	f (cm²)	(no limit on thickness)
Complete	≤ 15 %	≤ 9	12	0,10 A ku A ≤ 900 cm²	
Half full	15 % - 45 %	≤ 12	20	≤ 0,15 A ku A > 900 cm²	
With holes	45 % - 55 %	≤ 15	25	The minimum thickness indicated applies to a wall with natural elements. Specific assessments must be made for existing masonry.	
Carved stone masonry	–	–	24		
Irregular stone masonry, with brick bands	–	–	40		
Irregular stone masonry.	–	–	50		

ϕ = F/A – percentage of holes
F – the total surface area of permeable and deep but non-permeable holes
A – gross surface area
f – the average cross-sectional area of a hole

TAB. 3.4 (WALL. 4/A) - ASSESSMENT OF COMPRESSIVE RESISTANCE fk OF MASONRY [2]

Characteristic compressive strength fbk (*) of the element N/mm²	Full and 50 % full artificial elements				Natural elements with carved stones			
	M 15	M 10	M 5	M 2,5	M 15	M 10	M 5	M 2,5
2,0	1,2	1,2	1,2	1,2	1,0	1,0	1,0	1,0
3,0	2,2	2,2	2,2	2,0	2,2	2,2	2,2	2,0
5,0	3,5	3,4	3,3	3,0	3,5	3,4	3,3	3,0
7,5	5,0	4,5	4,1	3,5	5,0	4,5	4,1	3,5
10,0	6,2	5,3	4,7	4,1	6,2	5,3	4,7	4,1
15,0	8,2	6,7	6,0	5,1	8,2	6,7	6,0	5,1
20,0	9,7	8,0	7,0	6,1	9,7	8,0	7,0	6,1
30,0	12,0	10,0	8,6	7,2	12,0	10,0	8,6	7,2
40,0	14,3	12,0	10,4	-	14,3	12,0	10,4	-

(*)The characteristic compressive strength fbk is declared by the manufacturer of the elements according to EC standards.

Tab. 3.5 - TECHNICAL AND MECHANICAL CHARACTERISTICS OF WALL ELEMENTS [2]

WALL TABLE 1								
CLASSIFICATION AND AVERAGE COMPRESSIVE STRENGTHS FOR MORTAR [2]								
CLASS	RESISTANCE (1)		Type	Ingredients				
	kg/cm2	N/mm2		cement	aerated lime	hydraulic lime	sand	pozzolan
M1	120	12	with cement	1	-	-	3	-
M2	80	8	with cement	1	-	0,5	4	-
M3	50	5	bastard	1	-	1	5	-
M4	25	2,5	bastard	1	-	2	9	-
M4	25	2,5	pozzolan	-	1	-	-	3
M4	25	2,5	hydraulic	-	-	1	3	-
Mortars with different proportions in composition, mixed with additives, previously experimented, can be considered equivalent to those indicated when their average compressive strength results are not less than the values in the table..								

TABLE 3.6 (WALL 2)					
MECHANICAL CHARACTERISTICS OF NATURAL RESISTANT ELEMENTS					
Rock	Average value at rupture				**Ec**
	pressure		bending		
	N/mm2	kg/cm2	N/mm2	kg/cm2	N/mm2 * 10^3
granite/syenite	100-230	1000-2300	10 ÷20	100 ÷200	50 ÷ 60
porphyry	100-280	1000-2800	15 ÷ 22	150 ÷ 220	55 ÷ 65
sandstone	40-130	400-1300	4 ÷ 10	40 ÷ 100	10 ÷ 30
trevetin	20-60	200-600	5 ÷ 10	50 ÷ 100	8 ÷ 25
compact limestone	80-180	800-1800	6 ÷ 20	60 ÷ 200	45 ÷ 70
tuff	5 ÷ 15	50÷ 150	0,5 ÷ 0,7	5 ÷ 7	3 ÷ 15
marble	100-140	1000-1400	7 ÷ 20	70 ÷ 200	50 ÷ 80

Tab. 3.7 (WALL. 3) – Geometric requirements of earthquake-resistant walls [2]

Constructive typology	**tmin**	**(λ=ho/t)** max	**(l/h')** **min**
Ordinary masonry, made with square stone elements	300 mm	10	0,5
Ordinary masonry, made with artificial elements	240 mm	12	0,4
Reinforced masonry, made with artificial elements	240 mm	15	Cfaredo
Ordinary masonry, made with square stone elements, falling in the areas of grade 3 and 4	240 mm	12	0,3
Masonry made with semi-full artificial elements, falling in the area of grade 4	200 mm	20	0,3
Masonry made with full artificial elements, falling in the area of grade 4	150 mm	20	0,3

h' maximum height of the openings that limit the wall
l length of the wall
The value of the coefficient of flexibility λ must not exceed the value 20.

Tab. 3.8 (WALL. 4)- Surface of resistant masonry in each orthogonal direction for simple constructions **[2]**

Acceleration on the top of the terrain ag·S		≤0,07 g	≤0,1 g	≤0,15 g	≤0,20 g	≤0,25 g	≤0,30 g	≤0,35 g	≤0,40 g	≤0,45 g	≤0,4725 g
Type of structure	Number of floors										
Ordinary masonry	1	3,5 %	3,5 %	4,0 %	4,5 %	5,0 %	5,5 %	6,0 %	6,0 %	6,0 %	6,5 %
	2	4,0 %	4,0 %	4,5 %	5,0 %	5,5 %	6,0 %	6,5 %	6,5 %	6,5 %	7,0 %
	3	4,5 %	4,5 %	5,0 %	5,5 %	6,0 %	6,5 %	7,0 %			

Tab. 3.9 (WALL. 5A) - Characteristic shear strength in the absence of normal tension fvk0 (values in N/mm2) **[2]**

Type of supporting element	Elements characteristic compressive strength, fm	Mortar class	fvk0 (N/mm2)
Hollow and semi-hollow bricks	fbk> 15	M10 ≤ M ≤ M20	0.30
	7.5 < fbk< 15	M5 ≤ M ≤ M10	0.20
	fbk ≤ 15	M2.5 ≤ M ≤ M5	0.10
Reinforced concrete; calcium silicate; autoclaved concrete; natural stone Squared	fbk> 15	M10 ≤ M ≤ M20	0.20
	7.5 < fbk< 15	M10 ≤ M ≤ M20	0.15
	fbk ≤ 15	M10 ≤ M ≤ M20	0.10

CHAPTER 4 - THEORETICAL BASES [3]

4.1. Masonry Resistance
4.1.1. Basic Equations

Masonry is a heterogeneous material, as a result of its composition:
Solid or hollow blocks
Continuous mortar joints (beds)
Discontinuous or continuous joints (tops)
In general, we can call it a homogeneous material, anisotropic in terms of resistance and deformation. A masonry structure presents a behavior that is highly dependent on the reaction of its constituent elements, which are the Bricks and the Mortar.
- Brick is characterized by an elastic behavior, both in tension and in compression. With increasing baking time, bricks become more resistant, but more brittle and breakable.
- The behavior of the mortar depends heavily on the binder used and on the quantity.

4.1.2- MOHR-COULOMB THEORY

For highly axial loading conditions, the behavior of brick and mortar can be explained quite clearly by the Mohr-Coulomb criterion.

$$|\tau| \leq c - \sigma \cdot tg\phi \qquad (4.0)$$

c - is the cohesion
ϕ - is the angle of friction

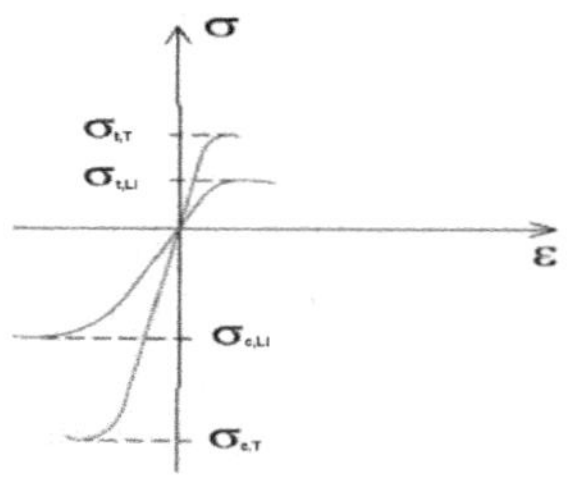

Fig. 4.1- Compression-tension stress diagram for mortar and brick

According to the MOHR-COULOMB theory, a mathematical model is created that describes the response of brittle materials such as concrete to shear and normal stresses. Most classical engineering materials follow this rule to some extent, at least in part, in the area of shear strength. Generally, this theory applies to materials for which the compressive strength greatly exceeds the tensile strength.

In geotechnical engineering, it is used to determine the shear strength of soils and rocks under various effective stresses.
In structural engineering, it is used to determine the failure load, as well as the angle of failure, caused by displacement in concrete and similar materials.
The hypothesis of Coulomb friction is used to determine the combination of shear and normal stresses that will cause a material to fail. The MOHR circle is used to determine which principal stresses will produce this combination of shear and normal stresses, and the plane angle at which this will occur.
According to the normality principle, the stress introduced at failure will be perpendicular to the line describing the failure mode.

It can be shown that a material that fails according to the Coulomb friction hypothesis will exhibit a displacement that precedes failure at an angle to the fracture line equal to the friction angle. This makes the resistance of the material determinable by comparing the external mechanical work, represented by the displacement and external load, with the internal mechanical work, represented by the deformation and stress at the fracture line.
According to the principle of conservation of energy, their sum must be zero, and this will make it possible to calculate the failure load of the structure.
A common improvement of this model is to combine the Coulomb friction hypothesis with the Rankine principal stress hypothesis to describe separation fracture.
The Mohr-Coulomb fracture criterion represents the linear envelope, which is taken from the shear strength diagram of a material, versus the applied normal stress.
This relationship is expressed as:

$$\tau = \sigma \ \mathbf{tang}\ (\varphi) + \mathbf{c} \qquad (4.1)$$

where
τ is the shear force
σ is the normal stress
c is the discontinuity of the shear stress, with the t axis, and
φ is the slope of the shear stress. The magnitude c is the cohesion and f is called the angle of internal friction.
The pressure is assumed to be positive in the following discussion. If the pressure is assumed to be negative then σ should be replaced by $-\sigma$.
If φ=**0**, the Mohr-Coulomb criterion is equivalent to the Tresca criterion.
On the other hand, if φ=900, the Mohr-Coulomb model is equivalent to the Rankine model. Higher values of φ are not allowed.
From the Mohr circle we have:

$$\sigma = \sigma\mathbf{m} - \tau\mathbf{m}\ \mathbf{sin}\ \varphi \qquad (4.2)$$

$$\tau = \tau\mathbf{m}\ \mathbf{cos}\ \varphi \qquad (4.3)$$

where $\tau m = (\sigma 1 - \sigma 3)/2 \quad \sigma\, m = (\sigma 1 + \sigma 3)/2$ (4.4) and σ_1 is the maximum principal stress, σ_3 is the minimum principal stress. Therefore, the Mohr-Coulomb criterion can also be expressed as:

$$\tau m = \sigma m \sin \varphi + c \cos \varphi$$

This form of the Mohr-Coulomb criterion is applicable to failure along a plane parallel to the σ_2 direction. It is easy to check that from the Mohr-Coulomb σ-τ relation we have:

$$(1)\ \sigma sh = \frac{2c \cdot \cos \phi}{1 - \sin \phi} \quad (4.5)$$

$$(2)\ \sigma t = \frac{2c \cdot \cos \phi}{1 + \sin \phi} \quad (4.6)$$

Considering that the Mohr – Coulomb circle has the equation:

$$(3)\ \left(\sigma - \frac{\sigma_1 + \sigma_2}{2}\right)^2 + \tau^2 = \left(\frac{\sigma_1 - \sigma_2}{2}\right)^2 \quad (4.7)$$

This equation is also represented by the color red in Mohr's circle.

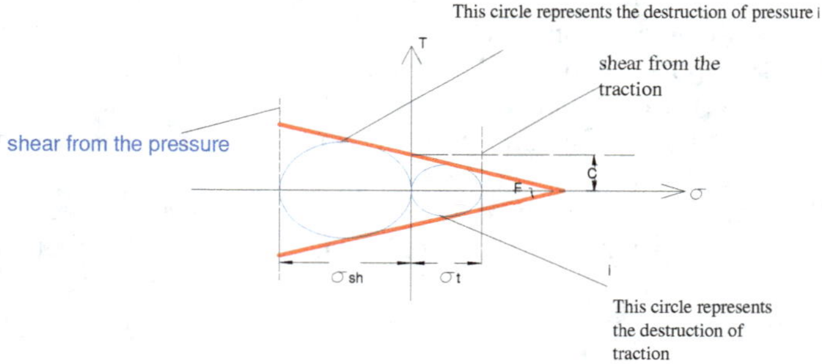

Fig. 4.2- Mori circle diagram for masonry

The same area can also be represented in the plane of the principal stresses $\sigma 1$ and $\sigma 2$

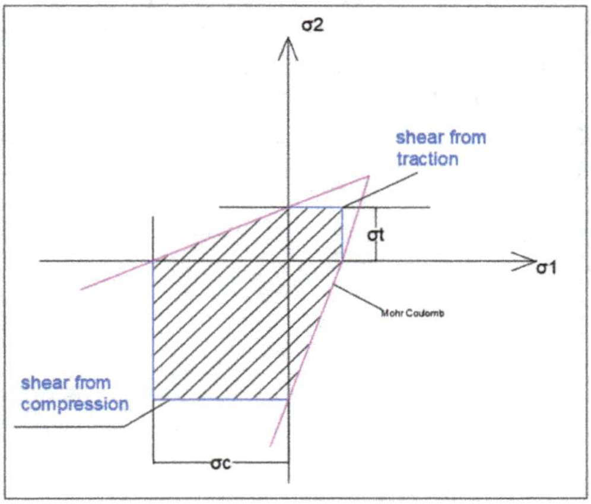

Fig. 4.3- Principal Stress Plan

In the tension and compression zone, the real behavior is actually between the Mohr-Coulomb criterion and the Galileo Maximum Stresses. In the biaxial compression zone, the field results wider.

The masonry material contains characteristics of its constituent elements but in its composition, it does not present an intermediate behavior between them, but some differences and specificities:

- First of all, the different deformability between the brick and the mortar, is responsible for the tense states of the multi-axial type, which make their behavior complex.
- Secondly, the presence of preferred directions of the material (direction of the mortar, holes in the bricks) give the masonry a very ORTHOTROPIC character, which means that the resistance does not depend only on the type of load applied, but also on its direction.

4.1.3- APPLICATION EXAMPLE [3]

To understand the complexity of the problem, we can study the case of a simple compression in a wall built from bricks and a layer of mortar, but only in the horizontal direction. The masonry material contains characteristics of its constituent elements, yet in its composition, it does not exhibit an intermediate behavior between them, but rather has distinct differences and specifics:

- First and foremost, the different deformability between the brick and the mortar is responsible for the state of stress of the highly axial type, which complicates their behavior.
- Secondly, the presence of preferred directions in the material (the direction of the mortar, the voids in the bricks) imparts a highly ORTHOTROPIC character to the masonry, meaning that its resistance depends not only on the type of load applied but also on the direction of that load. To understand the complexity of

the problem, we can study the case of simple compression in a masonry structure made of bricks and mortar layers only in the horizontal direction. We can assume the thickness of the mortar layer to be very small, allowing us to consider $\sigma_z^T = \sigma_z^{Ll} = 0$. With this assumption, we can accept that the masonry is subjected to a plane stress state.

The equilibrium of strains in the y direction:

$$\boldsymbol{\sigma_x^{Ll} + \gamma \cdot \sigma_z^T = 0} \text{ ku } \gamma = \frac{H}{h} \tag{4.8}$$

The equilibrium of strains in the x direction:

$$\varepsilon^T = \frac{1}{E^T}\left(\sigma_x^T - v^T \sigma_y^T\right) = \varepsilon^{Ll} = \frac{1}{E^{Ll}}\left(\sigma_x^{Ll} - v^{Ll}\sigma_y^{Ll}\right)$$

(4.9)

n- It's Poisson's ratio.

If we note: $\alpha = \frac{E^T}{E^{Ll}} \cdot \frac{v^{Ll}}{v^T}$ dhe $\beta = \frac{E^T}{E^{Ll}}$

After substitutions we will have: $\sigma_x^T = \frac{1-\alpha}{1+\beta\gamma} \cdot \sigma$ dhe $\sigma_x^{Ll} = -\gamma \cdot \frac{1-\alpha}{1+\beta\gamma} \cdot \sigma$

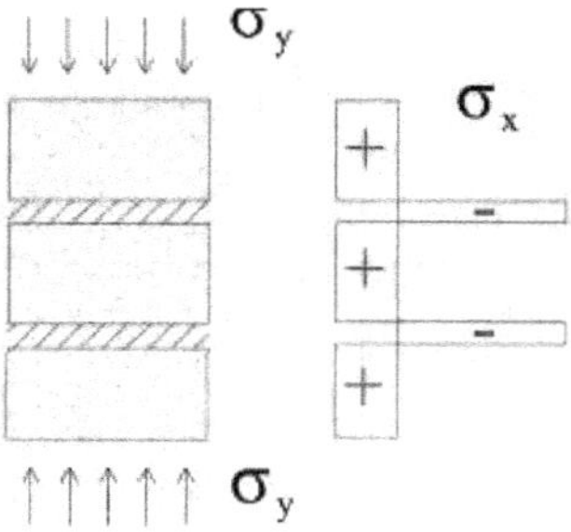

Taking the example $\alpha = 2\beta = 1.5$ and $\gamma = 10$

$\sigma_x^T = -\frac{1}{16}\sigma$ (Traction to the bricks)

Fig. 4.4- Stressed state of masonry

$\sigma_x^{Ll} = +\frac{1}{16}\sigma$ (Pressure in the mortar)

$\sigma_y^{Ll} = \sigma_y^T = \sigma$ (Pressure)

- The mortar results in more tension, but is subject to a triaxial compression. The presence of triaxial compression increases the resistance and avoids premature failure from instantaneous compression

We can generally accept that, in a wall composed of brick and mortar, failure and cracking occur in the bonding zone between the brick and the mortar. A significant increase in the separation zones between the brick and the mortar significantly reduces the resistance of the masonry.

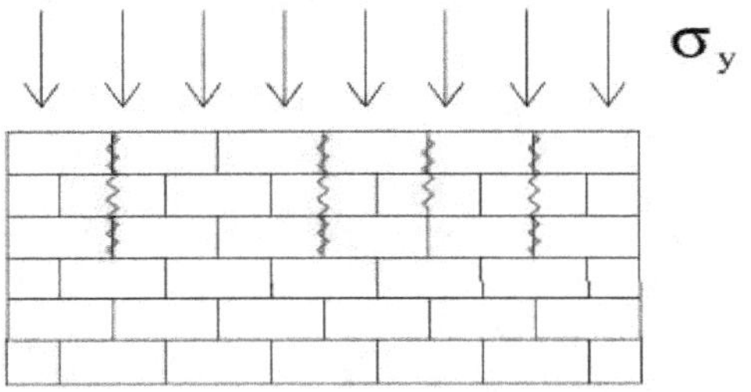

Fig. 4.5- Effects of load on masonry

If we do an experiment of a wall in simple compression, in general it is the brick that yields as a result of shear cracks, but this varies depending on the thickness of the mortar. The figure shows a typical example of a wall under the influence of simple compression. In this case, a derivation of the Galileo Maximum Stress criterion, also known as the Capurso–Sacchi Criterion, can be used, which considers a stressed state safe if the matrix:

$$\sigma + \sigma_R > 0 \qquad (4.10)$$

and: $$\sigma + \sigma_R \neq 0$$

where:

$[\sigma] = \begin{pmatrix} \sigma_{11} & \sigma_{12} \\ \sigma_{21} & \sigma_{22} \end{pmatrix}$ matrix representing the stress state in the masonry panel

$[\sigma_R] = \begin{pmatrix} \sigma_{R1} & 0 \\ 0 & \sigma_{R2} \end{pmatrix}$ matrix representing the compressive strength in two directions.

If we have: $det[\sigma + \sigma_R] = (\sigma_{11} + \sigma_{R1}) \cdot (\sigma_{22} + \sigma_{R2}) - (\sigma_{12} \cdot \sigma_{21}) = 0$ we will have the destruction of the panel.

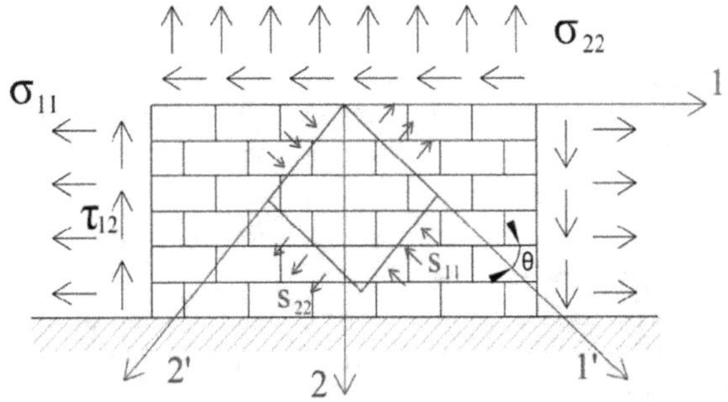

Fig. 4.6- Stressed state of plane masonry

Expressing this criterion in terms of the principal stresses, we will have:

$$\begin{aligned} \sigma_{11} &= s_1 \cdot cos^2\,\theta + s_2 \cdot sin^2\,\theta \\ \sigma_{12} &= (s_1 - s_2) \cdot cos\,\theta\, sin\,\theta \\ \sigma_{22} &= s_1 \cdot cos^2\,\theta + s_2 \cdot sin^2\,\theta \end{aligned} \tag{4.11}$$

Accepting we get the equation:

$$= (s_1 + \sigma_{R1}\, cos^2\,\theta + \sigma_{R2} \cdot sin^2\,\theta) \cdot (s_2 + \sigma_{R2}\, cos^2\,\theta + \sigma_{R1} \cdot sin^2\,\theta)$$
$$= (\sigma_{R2} - \sigma_{R1})\, sin^2 \cdot \theta\, cos^2\,\theta$$

In which $\sigma_{R1}<\sigma_{R2}$- compressive resistance, in the direction of the joint.
The above expression represents in each case the equation of a hyperbola, which passes through the point ($-\sigma_{R1}$, $-\sigma_{R2}$). This hyperbola is converted into a grid with θ=0 and θ=90°.

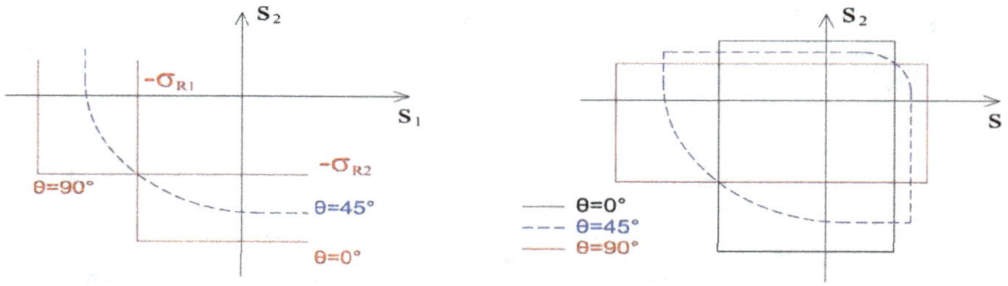

By developing an analogous criterion for shear, a whole resistant zone can be constructed, varying from the angle between the main directions and the 1st and 2nd directions of the panel. The resistant zone thus results as shown in the figure on the right.
In reality, the experimental zone (field) results different from the response of this model and in general results wider, especially in the compression-compression zone.
This is a consequence of:
- the underestimation caused by the model on the behavior of the individual constituent elements,
but above all:
- the fact that the real behavior from the influence of horizontal stresses, provoked by vertical loads, is not taken into account.
This fact can be understood from a simple example: a small lateral compression provokes almost a doubling of the vertical failure load. The results obtained from the experiments are illustrated in the figure below:

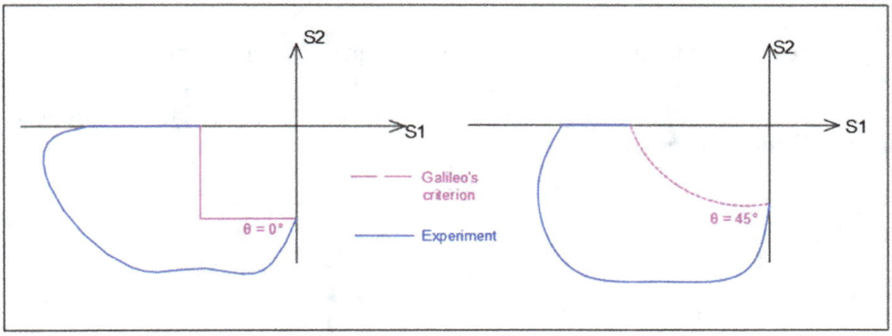

This also shows the importance of conducting experimental tests. Let us take the Mohr criterion back into consideration and extend it to the mass of masonry:

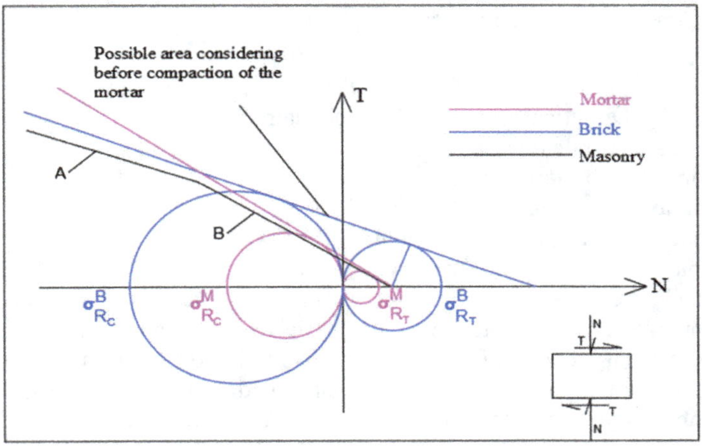

Fig. 4.7- A- Brick destruction B- Mortar destruction

his model appears to be very realistic. The destruction of the masonry comes as a result of the destruction of the weakest mass, namely the mortar.

4.1.4 EXPERIMENTAL TESTS [3]

Typical experimental tests performed on wall panels (macro-elements composed of 3 layers of bricks) are given in the figure below.
All four tests should be understood as conventional, while they initiate complex and difficult to cope with stress states in the panel.

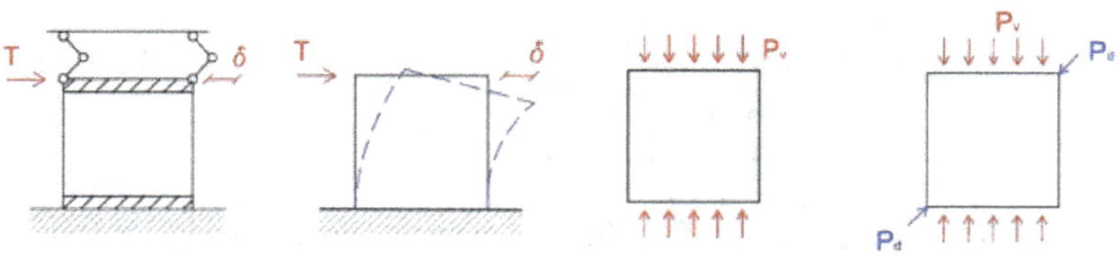

a b c d

Fig. 4.8- **a**- Shear test direct and bending- b- Compression test (the sides remain parallel)
c- Compression test d- simple diagonal

The difficulties of the theoretical type, related to the interpretation of experimental data, are correlated with:

- Non-uniform distribution of results
- Large variation of test shapes and dimensions
- Type of load (concentrated or distributed),
- Its direction
- Its mode of application (static or dynamic).

The diagonal compression test is taken as the basic test to express the shear characteristics of masonry, due to the simplicity of its implementation, as well as the possibility of knowing the stresses distributed within the panel.
The simple compression test shows the behavior of the panel under vertical loads.
To extend the results to a wider case of normal and shear stresses, in the panel are generally necessary suitable resistance criteria, able to show the response in all possible combinations.

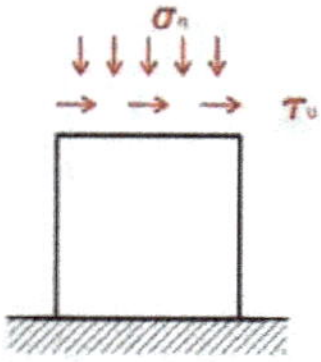

4.1.5 DESTRUCTION MECHANISMS

Below we present the possible mechanisms of collapse of a masonry mass:

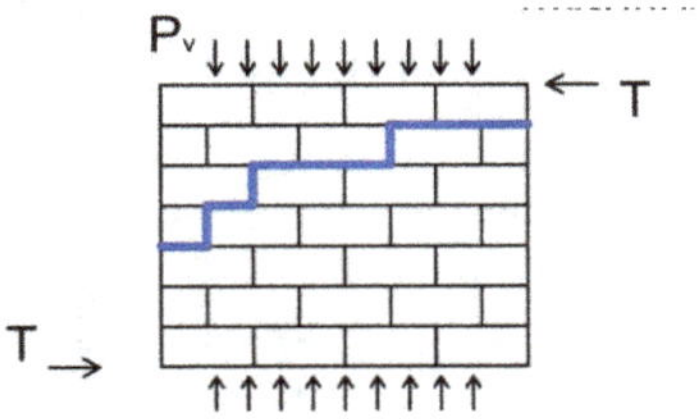

Slip of joints

This mode of failure is common in masonry made of resistant bricks, with relatively weak joints. In general, it is verified that the value of the P_v component is limited.

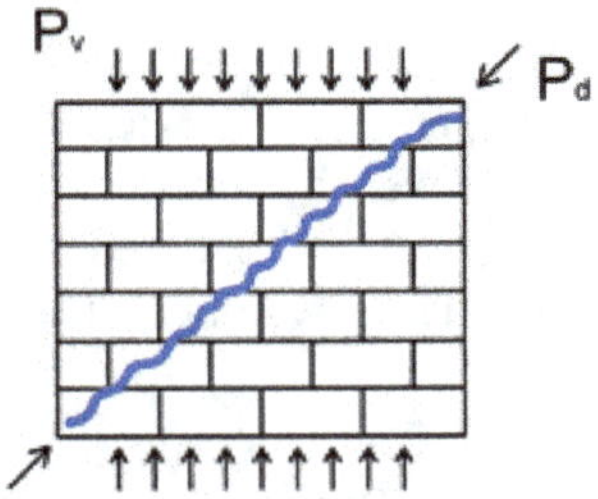

Break of the blocks This fracture is caused by the yielding of the blocks. Generally, includes the middle part of the masonry. The slope depends on the ratio between the vertical and horizontal components of the applied load. It is generally of interest to use hollow bricks with high-strength mortar binders.

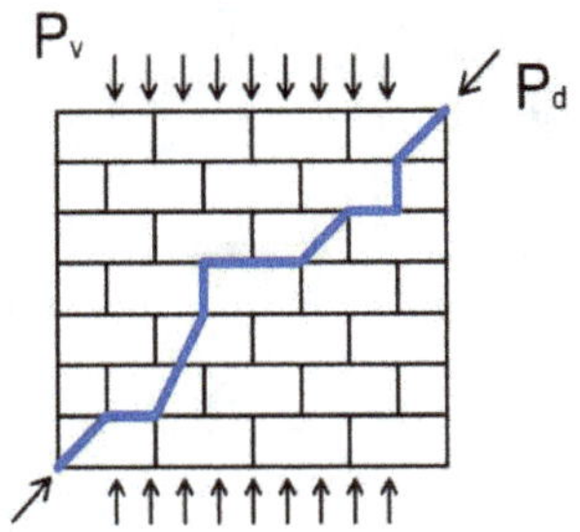

Combined mechanism

The crack generally follows the diagonal of the panel. This concerns both the mortar and the brick. It is a failure that generally occurs when the mortar and the brick have similar strengths. It is a type of failure that concerns masonry in general when there is a variety of loading combinations.

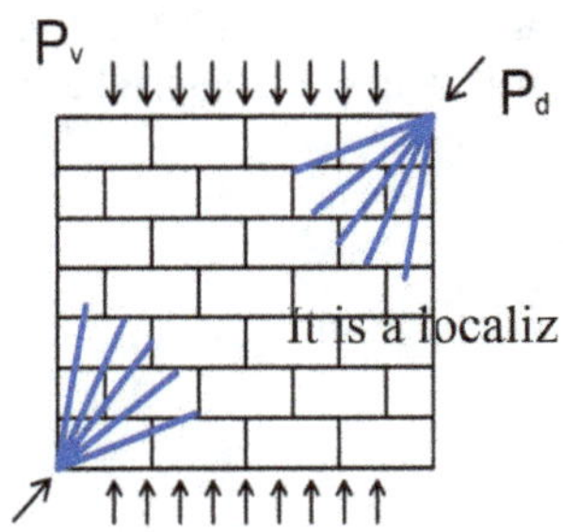

It is a localized failure, generally associated with a limited dimension of load application.

4.1.6 RESISTANCE CRITERIA

1 - <u>MOHR-COULOMB CRITERIA</u> [3]

As mentioned above, it's a criterion used by both Italian design norms and Eurocode 6, and provides:

$$\tau_u = \frac{T_u}{A} \tau_u = \tau_0 + \mu \sigma_n$$

(4.12)

where :

τ_0 - shear strength, when normal stresses are zero

μ - friction coefficient = 0.4

The values of τ_0 and μ are obtained from experimental tests on masonry panels, by varying the stress state on the panel itself.
This criterion cannot be taken as representative of the qualitative (destruction mode) and quantitative behavior of a given masonry. Moreover, it constitutes an estimate of the resistance of a given solid portion of the masonry, suitably reduced to take into account the distribution of experimental data. The advantage of this criterion lies in its simplicity.

2- CRITERIA based on reaching the failure state at the center of the panel

[2a] Yokel-Fattal Criterion

This criterion most often refers to the normal stresses in the panel:

$$\bar{\tau} = \frac{P_d}{b \cdot t\sqrt{2}} = \frac{T}{b \cdot t} \qquad (4.13)$$

$$\bar{\sigma}_y = -\frac{P_v}{b \cdot t} - \frac{P_d}{b \cdot t\sqrt{2}} = \frac{(P_v+N)}{b \cdot t} \qquad (4.14)$$

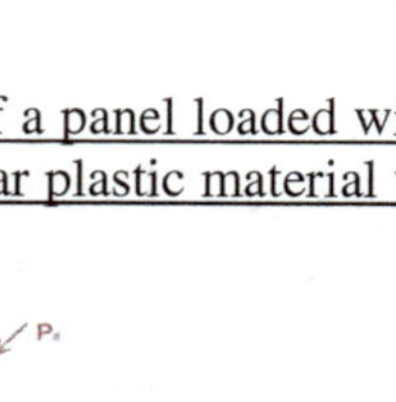

where t = panel thickness

<u>Froch has proposed a closed-form solution for the case of a panel loaded with a single diagonal load Pd, in the hypothetical case of a linear plastic material with *ν=0.2*</u>

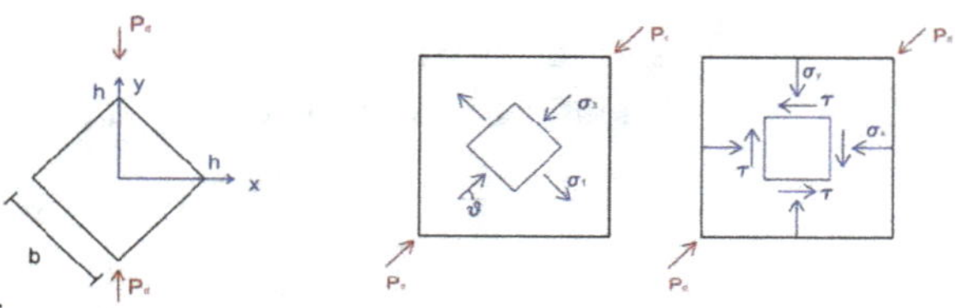

Where:

σ1 – Principal shear stress
σ3 – Principal compressive stress

In particular, in the center of the panel we would have

$$\sigma_1 = 0.734\bar{\tau}$$
$$\sigma_3 = -2.380\bar{\tau}$$

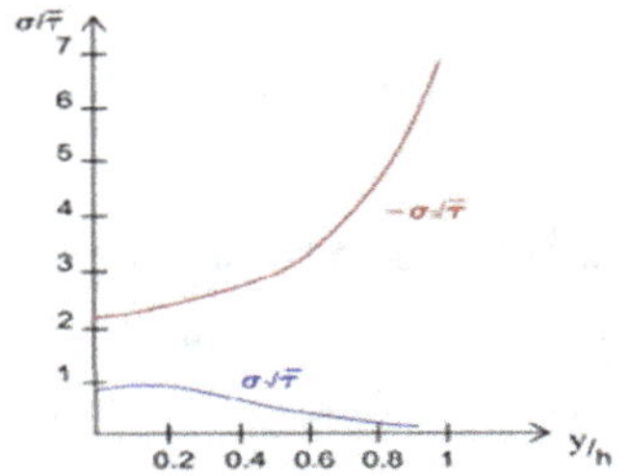

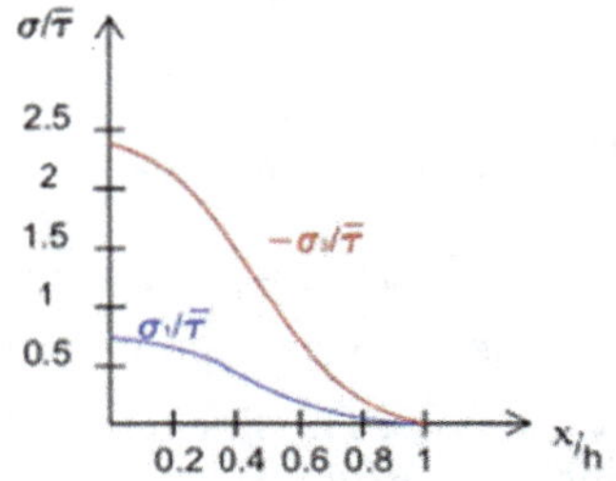

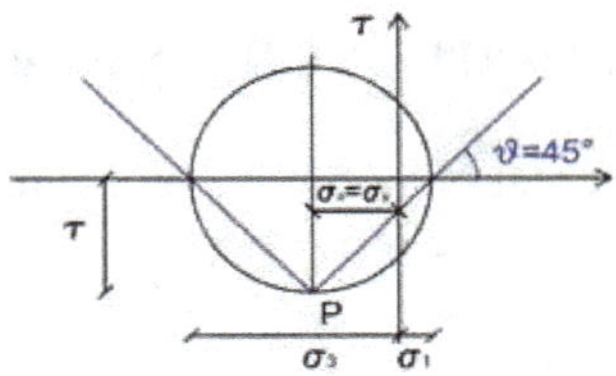

From such a theory, it is obtained that the distribution of stresses is not constant and in particular has an instantaneous variation, close to the unit value of the y/h ratio, as shown in the figure.

If we take into account the center point of the wall, it is indifferent to apply a stress s_1, σ_3, with $\theta = 45°$, or a stress case with $s_x = s_y$, reached in the Mohr circle. As a function of $\bar{\tau}$ in the panel we can say:

$$\sigma_1 = 0.734\bar{\tau}$$
$$\sigma_3 = -2.380\bar{\tau}$$

or

$$\sigma_x = \sigma_y = -0.823\bar{\tau}$$
$$\tau = 1.556\bar{\tau} \text{ (in pressure)}$$

Based on the diagonal test, failure can occur when $\bar{\tau} = \bar{\tau}_d$, being $\bar{\tau}_d$ the normal shear stress.
If a normal compressive stress Pv is applied to the panel, which causes a stress at the center of the panel:

$$\sigma_v = -\frac{P_v}{b \cdot t}$$

By superimposing the effects on the panel, the main stresses in it will be:

$$\sigma_1 = \frac{\sigma_x+\sigma_y}{2} + \frac{\sigma_v}{2} + \sqrt{\tau^2 + \left(\frac{\sigma_v}{2}\right)^2} = -0.823\bar{\tau} + \frac{\sigma_v}{2} + \sqrt{(1.556\bar{\tau})^2 + \left(\frac{\sigma_v}{2}\right)^2}$$
(a)

And shear failure curve:

$$\sigma_3 = \frac{\sigma_x+\sigma_y}{2} + \frac{\sigma_v}{2} - \sqrt{\tau^2 + \left(\frac{\sigma_v}{2}\right)^2} = -0.823\bar{\tau} + \frac{\sigma_v}{2} + \sqrt{(1.556\bar{\tau})^2 + \left(\frac{\sigma_v}{2}\right)^2}$$
(b)

$$\theta = \frac{1}{2} arctg\left(\frac{\sigma_v}{3.112\bar{\tau}}\right)$$

Under these conditions it is possible to determine the resistance as a function of $\frac{\bar{\tau}}{\bar{\tau}_d} dhe \frac{\bar{\sigma}_y}{\bar{\tau}_d}$, by adapting an appropriate resistance criterion, expressed in terms of the principal stresses. The simplest are the criteria of the critical normal stress and the critical elongation.

[2b] CRITERION of critical normal stresses or extreme stresses.

Destruction occurs when the maximum withstand voltage is reached at a point, which is defined as:

$$f_m = 0.734\ \bar{\tau}_d \quad \text{(simple shear)}$$
(4.18)

max shear stress in the center of the panel for Pd=Pd‘, Pv = 0, P’d = diagonal breaking load.

$$\bar{\tau}_d = \frac{P_d'}{bt\sqrt{2}} \qquad f_m' = -\frac{P_v'}{bt} \quad \text{(simple pressure)}$$

The simple compressive strength of the panel is when $\boldsymbol{P_v = P_v', P_d = 0}$. According to this criterion, for any combination of P_v and P_d, collapse is reached when:

$$\sigma_1 = \frac{\sigma_x+\sigma_y}{2} + \frac{\sigma_v}{2} + \sqrt{\tau^2 + \left(\frac{\sigma_v}{2}\right)^2} = fm \quad \text{(a') shear failure curve}$$

$$\sigma_3 = \frac{\sigma_x+\sigma_y}{2} + \frac{\sigma_v}{2} - \sqrt{\tau^2 + \left(\frac{\sigma_v}{2}\right)^2} = f'm \quad \text{(b') compression failure curve}$$

Normally it is (a') that conditions the crisis state, for low values of normal stresses..

2- TURNSEK-CACOVIC Model (based on the POR calculation method) [3]

This model is valid in the case of diagonal failure and is based on experimental tests of type b, that is, when the masonry element is subject to combined compression and shear loading.

The "pier" (wall) is taken into analysis, subject to a vertical load P_v and a horizontal load P_h, at the extreme part of the element, as shown in the figure. Under the hypothesis that the tangential stresses have a behavior according to the Jourawsky theory, we would have:

$$\sigma_v = \frac{P_v}{bt}; \qquad \bar{\tau}_{xy} = \frac{P_H}{bt}$$

Where $\bar{\tau}_{xy}$ is defined as the average shear stress or conventional stress.

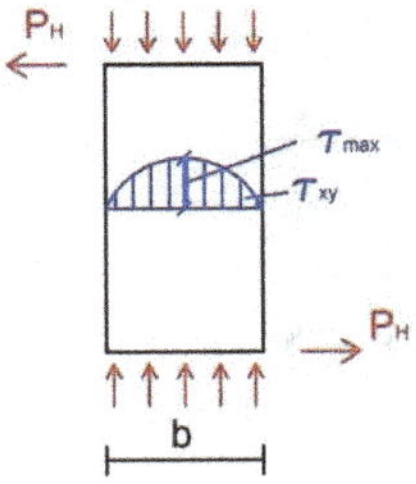

$$\sigma_{sh} = \frac{P_v}{bt}$$

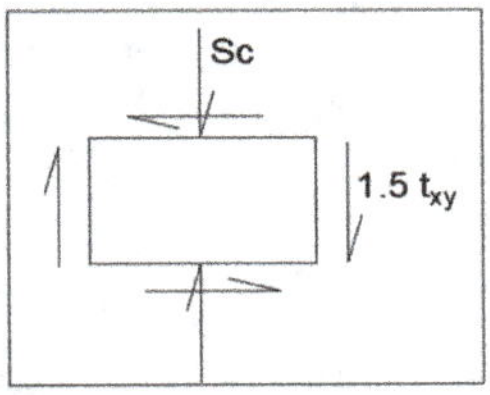

Analogously to the Yokel and Fattal criterion, the verification is performed at the center of the panel where it is assumed that the stressed state is characterized by:

$$\sigma sh, \tau_{xy,max}$$

The vertical compression stresses are : $\sigma_c = \sigma_v = \frac{P_v}{bt}$

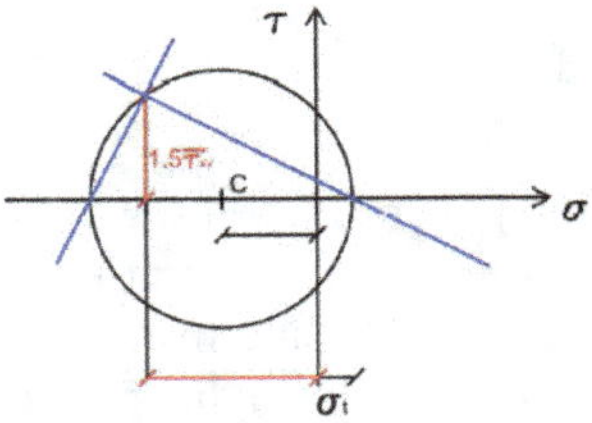

The maximum tangential stresses are:

$$\tau_{xy,max} = 1.5\bar{\tau}_{xy} = 1.5\frac{P_H}{b \cdot t}$$

The main tensile stresses result equal to:

$$\sigma t = -\frac{\sigma sh}{2} + \sqrt{(1.5\bar{\tau}_{xy})^2 + \left(\frac{\sigma_{sh}}{2}\right)^2} = f_m \qquad (4.19)$$

Destruction is achieved when $\sigma_t = f_m$

In the case of a clean shear (Pv=0, so $\sigma_{sh} = 0$ and with the Mohr circle centered at the origin) the destruction is achieved for $\sigma_t = 1.5\tau_{xy0}$,, i.e. $f_m=1.5\tau_{xy0}$. Substituting into the above relations we would have:

$$-\frac{\sigma_{sh}}{2} + \sqrt{(1.5\bar{\tau}_{xy})^2 + (\frac{\sigma_{sh}}{2})^2} = 1.5\tau_{xy0}$$

Under these conditions, the Turnsek-Cacovic criterion will be expressed by the following relation:

$$\bar{\tau}_{xy} = \tau_{xy_0}\sqrt{1 + \frac{\sigma_{sh}}{1.5\tau_{xy_0}}}$$

which for its simplicity is used in many normatives. If we set $\bar{\sigma}_y = -\sigma_{sh}$ and $\bar{\tau} = \bar{\tau}_{xy}$ we can make a comparison with the Yokel criterion. After a mathematical transition we can have:

$$\bar{\tau} = \frac{f_m}{1.5}\sqrt{1 - \frac{\bar{\sigma}_y}{1.5 f_m}}$$

(4.20)

From the comparison it is observed that the Turnsek criterion should generally be used with great caution in the field of high compressive stresses.

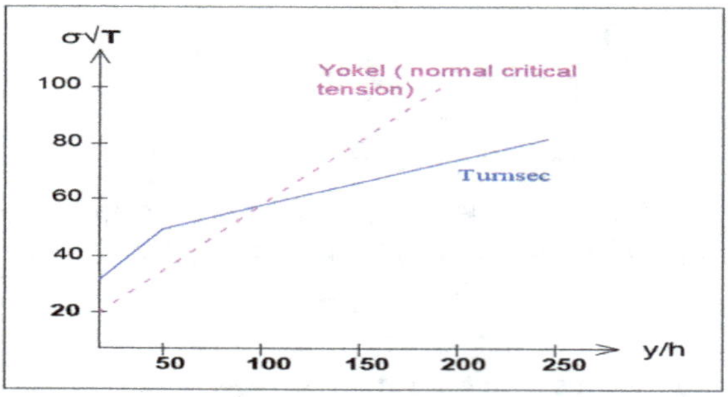

In general, when several panels are subjected to the test, the following experimental response is obtained:
1 and **3**- failure from rotation of the joints (mortar)
2- failure from diagonal cracks
4- failure from breakage of bricks from bending with compression.
Zone 2 in most cases can be increased even for low values of $\bar{\sigma}$ if high quality bricks are not used.

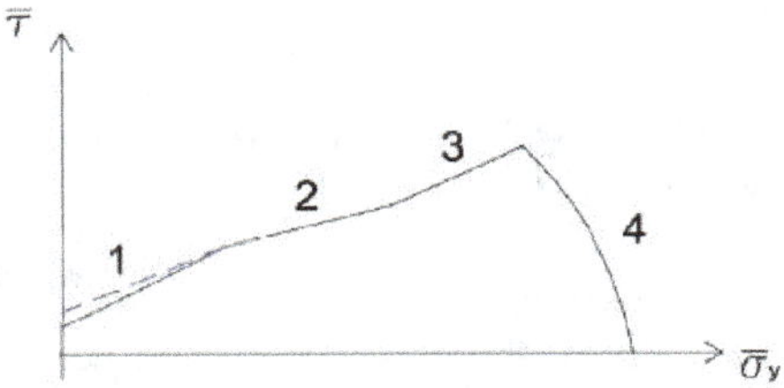

The Mohr-Coulomb criterion is in favor of safety, while its coefficient μ =0.4 is calibrated for the first three parts of the graph.
THE STANDARDS PROVIDE THE FOLLOWING VERIFICATION FORMULAS:

Destructions from joint slippage (of the mortar):

$V_{Rd} = \frac{1}{\gamma_m} t(\tau_{ok} + \mu\sigma_n)$ [kN/m] (value per unit length, b=1)

where:
τ_{0k} = characteristic value of surface tension
$\mu \cong 0.4$
σ_n= average compressive stresses

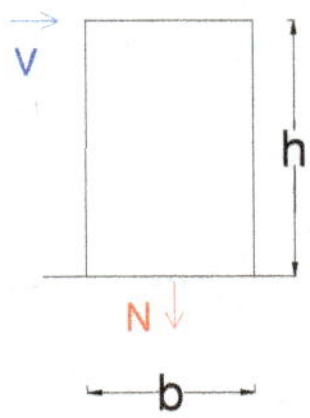

Failure due to diagonal cracking (modified Turnsek equation)

$V_{Rd} = \frac{1}{\gamma_m} (t \frac{f_{tk}}{P} \sqrt{1 + \frac{\sigma_n}{f_{tk}}})$ [kN/m] (value per unit length, b=1)

Where: P depends on the distribution of τ in the panel:
P = 1.0 (constant distribution in the panel)
P = 1.5 (parabolic distribution)
In general P=1.5, although values of P=1 come closer to the experimental behavior

$f_{tk} = 0.5 \frac{V}{(\frac{b+h}{2})t}$ = characteristic value in the case of diagonal tests.
The coefficient 0.4 is often achieved based on experimental evidence.

Destruction by pressure

Destruction by compression is generally avoided by limiting the compressive stresses in the panel. Alternatively, the Yokel criterion can be adopted, which, as we have noted previously, is sufficiently in favor of resistance. If the verifications are not satisfactory, one can:

- to increase the value of f_{tk} and τ_{0k}, improving the properties of the mortar,

- by applying a compressive stress σ_c.

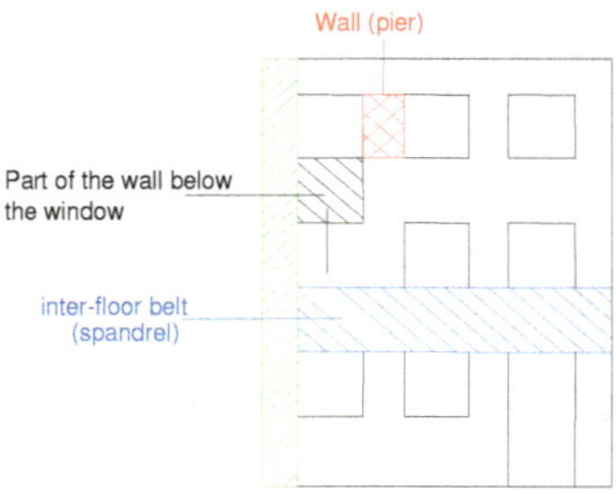

4.1.7. RESISTANT MECHANISM of a shear wall (or bracing) [3]

They are defined as:
"pier"- part of the masonry located between two windows, which run parallel from floor to floor.
"spandrel"- continuous strip located between the window of the floor above and the window of the floor below, on each floor. If the wall in section represents a sufficiently rigid strip of the floor, such that To ensure the parallel functioning of the wall elements, it is possible to individualize at each level a resistant mechanism, in which the elementary cell is the basement.
In these hypotheses, by analyzing a single "cell", we can observe that the collapse comes from or is caused by the breakage of the "pier", with a mechanism characterized by the formation of plastic hinges in them.

Fig. 4.9 - Deformation of piers due to shear force transmitted from spandrel

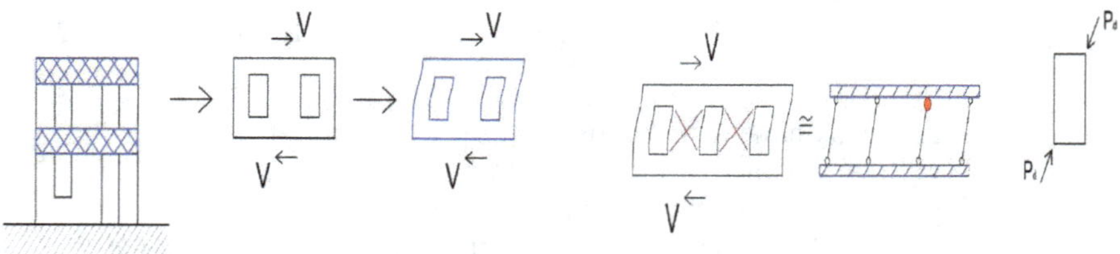

In this case, each "pier" results subject to a stressed state (shear stress), comparable to that reproduced in a diagonal test, therefore the criteria explained previously can be applied.
The collapse of the wall is reached when the first fracture of the "pier" is achieved.

4.2. **POR METHOD**

The POR method is a verification method that assumes the following calculation hypotheses:
- masonry structure with box-type resistant mechanism
- infinitely rigid slab in its own plane

• perfectly connected walls and slabs
• walls in which the floor "bands" result rigid
• destruction of the "**walls**" mainly in shear with an angle of 45°
• elasto-plastic behavior of the "**walls**".
• the resistance of other elements from the "**walls**" is neglected.

The POR method accepts as a behavior model that of the PANEL proposed by Turnsek-Cacovic, based on which:

$$T_u = A\tau_k\sqrt{1+\frac{\sigma_0}{1.5\tau_k}}$$

(4.21)

where:

T_u - maximum shear force absorbable by the wall, (limiting resistance of the panel)

τ_k - shear resistance of the masonry.

σ_0 - normal stresses due to applied vertical loads.

In the absence of experimental data, for verification with the POR method the following values are accepted:

$$E = 6 * G\,; \quad G = 1100\tau_k[{}^{t}/_{m^2}]$$

An elasto-plastic behavior is assumed for the panel and the following static scheme is considered:

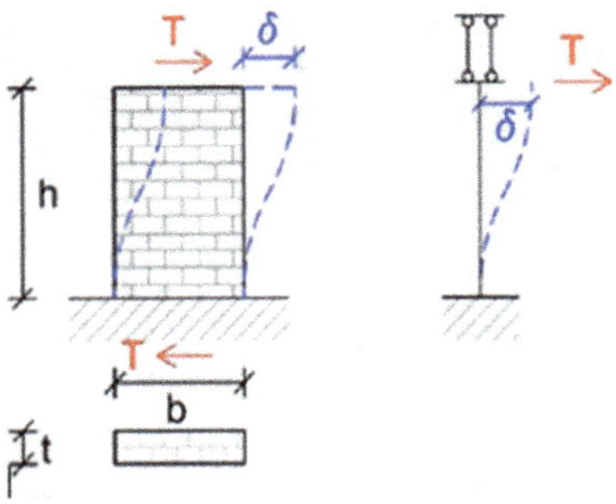

$$\delta = \delta_T + \delta_F = T\frac{h}{GA/\chi} + \frac{T\cdot h^3}{12EJ} \qquad J = \frac{t\cdot b^3}{12} = A\frac{b^2}{12}$$

$$T = \frac{GA}{\chi\cdot h}\left(\frac{1}{1+\frac{GA}{\chi}\frac{h^2}{12EJ}}\right)\delta = k_o\delta$$

Where:

χ - masonry ductility

T- elastic stiffness in bending

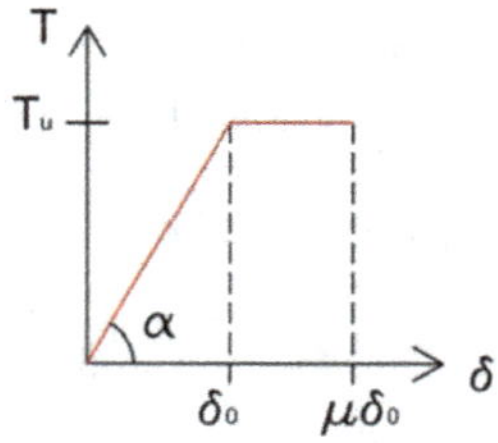

shear for the mural panel

$$tg\alpha = k_0 = \frac{GA}{1.2h}\frac{1}{1+\frac{1}{1.2}\frac{G}{E}\left(\frac{h}{b}\right)^2}$$

4.2.1. EXAMPLE 1

The shear characteristics for the following pier (wall) are required to be determined:

Data:		Loads:	
Length:	b = 1.30 m	Vertical stresses:	σ_0= 5 t/m^2
Thickness:	t = 0.50 m		σ_0= 50.0 kN/m^2
Height:	h = 2.50 m		

Masonry with injected stones:

τ_k = 11 t/m^2 = 110 kN/m^2
σ_k= 300 t/m^2 = 3.000 kN/ m^2 = 3 N/ mm^2
G = 121.00 t/m^2 = 1.21x10^5 kN/m^2
E = 6 G = 7.26 x10^6 kN/m^2

Boundary resistance:

$$T_u = A\tau_k\sqrt{1+\frac{\sigma_0}{1.5\tau_k}} = 1.30 \times 0.50 \times 110\sqrt{1+\frac{50}{1.5 \times 110}} = 81.6kN$$

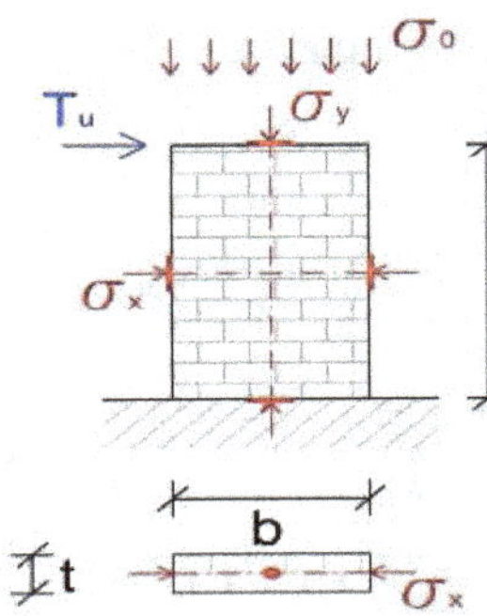

Total elastic stiffness:

$$k_0 = tg\alpha = \frac{GA}{1.2h}\frac{1}{1+\frac{1}{1.2}\frac{G}{E}\left(\frac{h}{b}\right)^2} = \frac{1.21\times10^5\times1.30\times0.5}{1.2\times2.50}\frac{1}{1+\frac{1}{1.2}\frac{1}{6}\left(\frac{2.50}{1.30}\right)^2} = 17320kN$$

Displacements at the elastic limit: $\delta_0 = \frac{T_u}{k_0} = \frac{81.60}{17320} = 4.7\times10^{-3}m = 0.47$cm

Limit displacement (ductility $\mu = 1.5$) : $\boldsymbol{\delta_u = \mu\delta_0 = 1.5\times 0.47cm} = 0.70$ cm

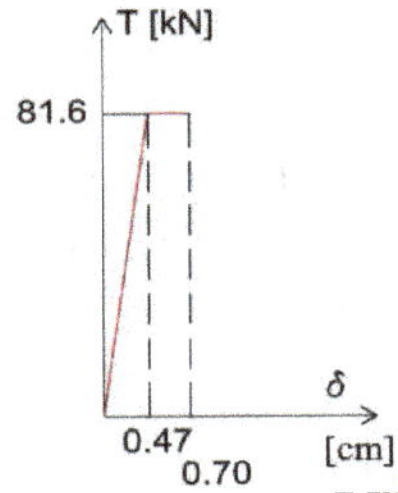

g. Static scheme, elastic and plastic displacement

<u>The POR method</u> predicts that the mural wall can be schematized as a set of walls (piers) that work in parallel. Each element is subjected to the same relative displacements between the upper part and the embedment in the base, that is, between the two slabs that are considered rigid.

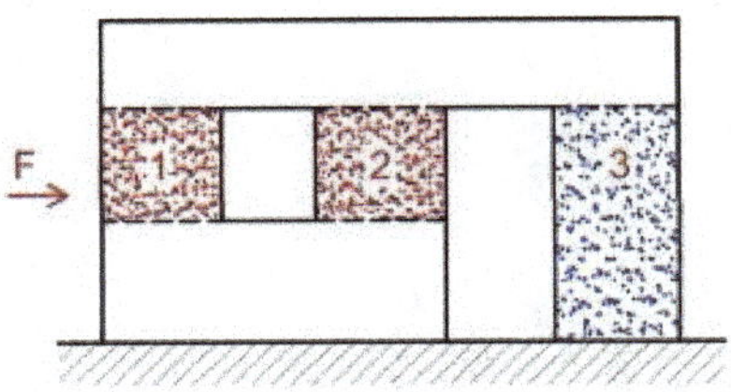

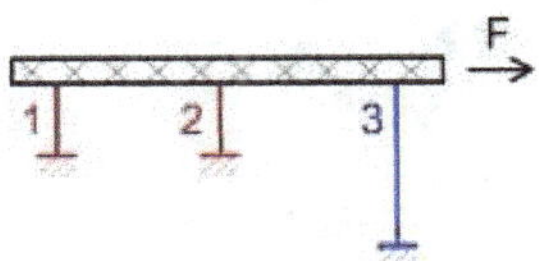

Fig. 4.11- Schematization of a wall facade in pier and spandrel elements

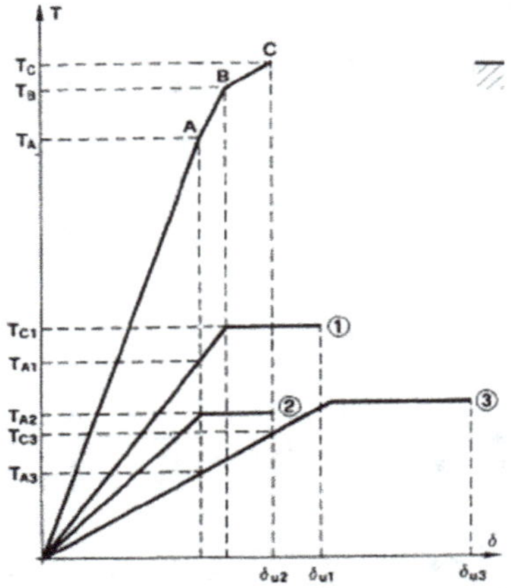

The behavior of the i-th floor is obtained by summing each displacement contribution of each wall. The process of accumulation of these contributions ends when one of the piers reaches the ultimate displacement (determined in function of the wall ductility). The sum of the resisting contributions for such a displacement determines the maximum reactive force $\mathbf{F_R}$, or the developed resistance of the i-th floor.

At point C, the wall breaks due to the failure of slope no. 2.

In the case of a multi-story building, the check is made for each floor. The forces of this floor are divided among the walls according to their stiffness, determined as above. For each floor, it must be checked that the acting force F_S is less than or equal to the resisting force F_R:

$$F_S \leq F_R \tag{4.22}$$

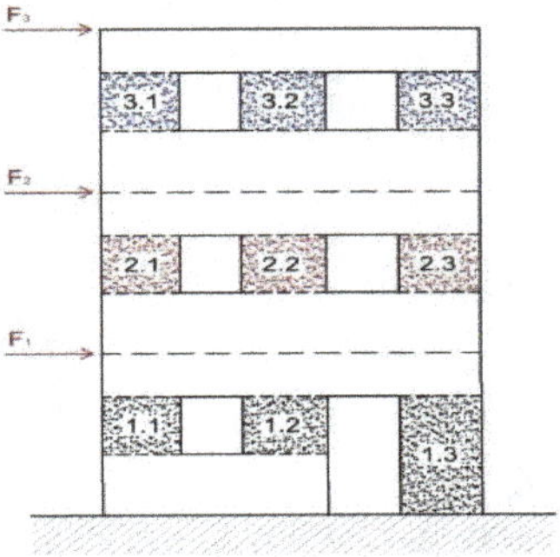

There are no interactions between floors

Fig. 4.12- Discretization of the wall into elements

Furthermore, by comparing the reaction forces of different floors, the weakest floor can be determined. Usually, we do not limit ourselves to a check at the elastic limit but calculate up to the collapse state when a wall exceeds the value of the displacement at the elastic limit, the behavior of the system becomes nonlinear, and there is no longer a proportionality between the forces and the displacements. In these cases, it is necessary to proceed with a nonlinear calculation:

-

-The position of the center of gravity is calculated as a function of the vertical pressures acting on each pier.
-The position of the center of rigidity is calculated.
-The value of the elastic stiffness resistance and the elastic limit displacement of each individual wall is calculated.
-Displacement increments are applied according to the x (or y) direction of the center of mass, adjusting the stiffness of the walls in the plastic zone and the position of the center of rigidity until collapse.
In the most general case, in which horizontal or vertical ties are present (prestressing action in each panel), the formula proposed by Turnsec-Cacovic is modified:

$$T_u = A\tau_k\sqrt{1 + \frac{\sigma_0+\sigma_x+\sigma_y}{1.5\tau_k} + \frac{(\sigma_0+\sigma_y)\cdot\sigma_x}{2.25\tau_k^2}} \tag{4.23}$$

where:
Tu = ultimate shear force absorbed by the panel = ultimate resistance of the panel.

σ_x = normal horizontal stress at the center of the panel due to the horizontal braces (horizontal prestressing)
σ_y = vertical stress at the center of the panel due to the vertical braces (vertical horizontal)

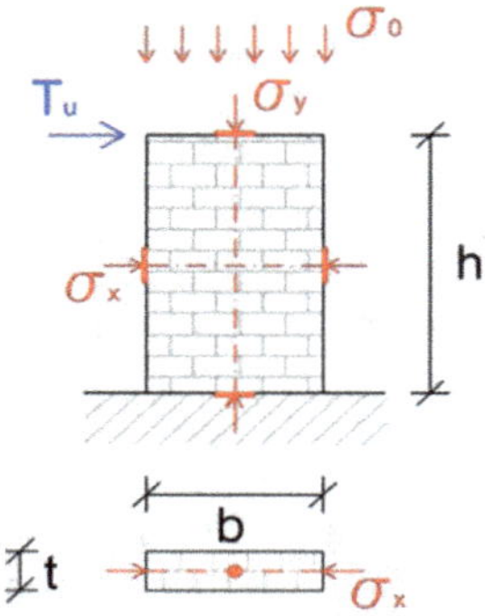

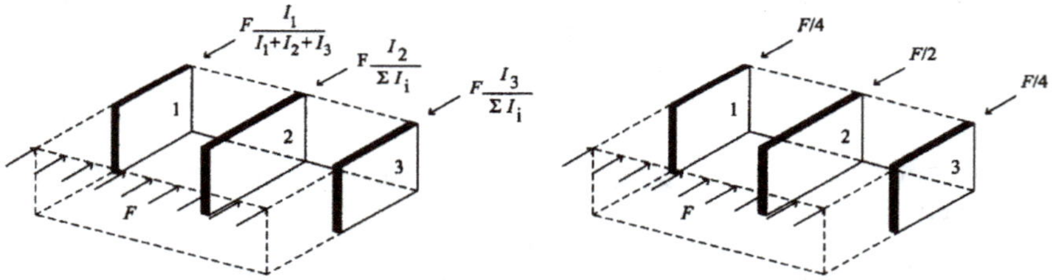

Fig. 4.13 - Transmission of horizontal floor force to load-bearing walls

Some considerations on the POR method:

1. The assessment assumes that the building's collapse mechanism is of the "shear" type. This method is valid only if the building indeed exhibits a collapse mechanism characterized by cracking at an angle of 45 degrees to the horizontal forces (such as those from an earthquake). This scenario is more plausible for heavy, short buildings with significant amounts of reactive masonry, such as older stone masonry structures. In contrast, it is less likely to apply to newer buildings with thin brick load-bearing masonry, predominantly subjected to compression and flexural stresses.
2. All walls are assumed to be embedded at the edges. This hypothesis assumes that the floor beams are rigid and this is true for short and old-designed buildings. Today, due to the reduced thickness of the load-bearing walls, the reduced height of the mezzanine, the part below and above the window can be called a simple "veil", practically non-structural. In this latter case it would be more realistic to consider walls with a height equal to the height of the mezzanine,

embedded at the base and connected at the top by connecting rods, which serve more or less as a rigid crossbar.

3. The method, based on the Turnsek-Cacovic criterion, considers the characteristic shear strength τ_k as the value that refers to the first crack in the center of the panel, in a diagonal test. The value τ_k is obtained by assuming that we have a homogeneous and perfectly elastic material until cracking, with a precise distribution of normal and tangential stresses in the center of the panel.
4. The method considers only elements made of masonry (walls) but does not consider the possibility of consolidating the structure with elements of different materials and typologies, e.g. reinforced concrete walls or metal structures with bracing.

5. The method does not in any way control eventual crises due to:
- Collapse of the floor beams, assumed to be infinitely rigid
- Lack of connection between the slab and the masonry
- Separation between orthogonal walls
- Inadequate rigidity of the floors
- Destruction of the walls due to exceeding the maximum compressive stresses

N.B. The POR method is a method generally used for adapting existing structures, after verifying the assumptions that are based on the method itself.

4.3. METHOD OF EQUIVALENT FRAMEWORKS.

For modern buildings, or for those with floor beams that exhibit slight rigidity, an "elasticity" check should be performed by simulating the walls as a frame system. This method involves modeling masonry buildings using "macroelements". In plan modeling, each wall is treated individually, while three-dimensional modeling takes into account the entire building structure.

In this numerical presentation, the structure is schematized as an equivalent frame composed of:
- Elements on the vertical axis: ordinary masonry panels or b/a, columns or b/a sets
- Elements on the horizontal axis: connecting beams, floor beams, b/a connecting strips.
- This schematization is acceptable in the case when the geometry of the walls and the distribution of openings in them are characterized by a certain regularity, especially with regard to the order of the openings. Each element of the frame, with horizontal or vertical development, is represented as a "one-dimensional" element through its main barycentric axis.

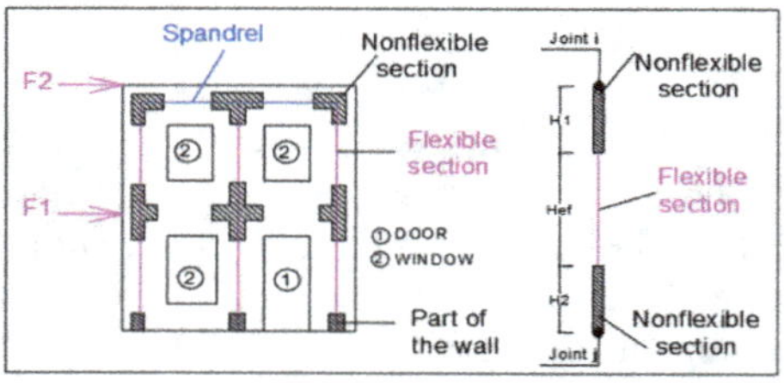

Fig. 4.14- Schematization of a wall facade in equivalent frames

- The introduction of infinitely rigid sections, with appropriate dimensions at the ends of the elements (piers, girders), allows to model the reduced deformation of the masonry areas, limited by openings in the walls (structural joints).

The height of the deformable part, or the effective height of the Heff spandrel, must be determined in such a way as to take into account approximately the deformability of the masonry in the joint areas; it can be estimated based on the following relationship, proposed by Dolce (1989), as a function of the geometric dimensions of the panel and the openings:

$$H_{eff} = h' + \frac{l \cdot (H - h')}{3 \cdot h'} \leq H$$

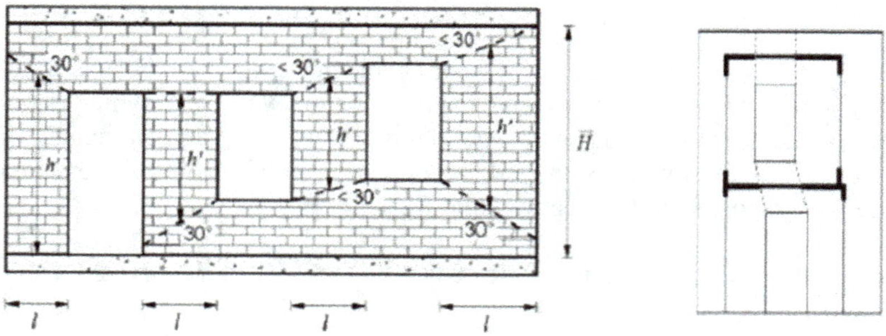

Fig. 4.15- Definition and deformed scheme of equivalent frames

where:

l – is the length of the panel

h- is the net height of the floor

h' – is a conventional height parameter that is determined based on the case studies reported in the figure.

After elastic analysis has been performed on the space frame system of the structure, which is subjected to a distribution of horizontal forces (seismic or wind), each wall (pier) is found to be subject to N, M, T actions.

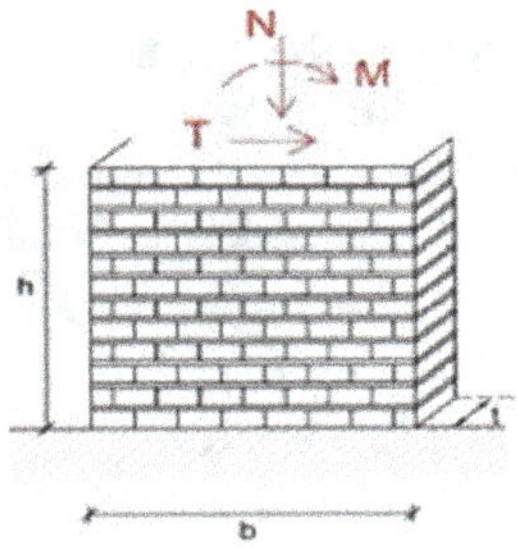

Fig. 4.16- Forces acting on a pier, wall panel

4.4. DESIGNS IN NON-SEISMIC AREAS [3]

4.4.1. Computational operations.

Veprimet mbi konstruksionet grupohen ne menyre qe te percaktojne kushtet e ngarkeses, te tilla qe te rezultojne me te disfavorshmet per qellim te kontrolleve te vecanta, duke patur parasysh mundesine e reduktuar te pjesemarrjes se njekohshme te gjitha veprimeve, me vlerat me te pafavorshme respektive, sic lejohen nga normat ne fuqi. Veprimet e projektit jane ato te ekspozuara ne forme te pergjithshme ne paragrafin 3.4, dhe raportuar ketu ne formulimin me te perdorshem :

$$F_d \gamma_S G_k + \gamma_q G_{1k} + \sum_{i=2}^{n} \gamma_q \psi_i G_{ik} \qquad (4.25)$$

where:
Gk – permanent loads
Q_{ik} – temporary loads
γ_s – partial safety factor which is:
1.3(1.5 – 1.0) for ultimate limit state control.
1.0 for operational limit state control
γ_q –partial safety factor which is:
1.5(0.0) for ultimate limit state control.
1.0 for operational limit state control
ψi – combination factor for different scenarios (ULS/ELS).
ψ_i – koeficient I kombinimit per skenare te ndryshme (ULS/ELS).

4.4.2. Calculation RESISTANCE and modulus of elasticity.

EC and Italian standards prescribe that the characteristic compressive strength f_{bk} of the masonry must be determined experimentally. However, an exception is accepted in the case of walls made of artificial elements, solid or semi-solid (or natural elements), with horizontal and vertical mortar joints. In this case,

observing the resistance characteristics of the elements and the mortar, the compressive strength is concluded. (see TABLE WALL. 5/A), Chapter Tables.
In the case of masonry made of natural elements, the characteristic compressive strength of the element is conventionally assumed to be equal to

$$f_{bk} = 0.75 f_{bm} \tag{4.26}$$

where:
F_{bm} – is the average compressive strength of squared stone elements.
The characteristic shear strength **f_{vk}** in the presence of compressive stresses is determined:

$$f_{vk} = f_{vk0} + 0.4\sigma_n \tag{4.27}$$

where:
fvk0 – is the shear resistance, in the absence of normal stresses.
σ_n – is the normal stress, caused by vertical loads acting on the verification section.
In the case of artificial elements, half-full, or with holes, the stress should be:

$$\mathbf{f_{vk} \leq f_{vk,lim} = 1.4 * f_{bk}}$$

In the absence of experimental tests, the shear resistance of masonry f_{vkO}, when normal stresses are absent, can be derived from the compressive resistance f_{bk}, according to the table below.

Tab. 4.1. Characteristic shear strength in the absence of normal tension fvk0 (values in N/mm2)

Type of supporting element	Characteristic compressive strength fm of the element	Mortar class	fvk0 (N/mm2)
Empty and half-empty bricks	fbk> 15	M10 ≤ M ≤ M20	0.30
	7.5 < fbk< 15	M5 ≤ M ≤ M10	0.20
	fbk ≤ 15	M2.5 ≤ M ≤ M5	0.10
Reinforced concrete; calcium silicate; autoclaved concrete; squared natural stone	fbk> 15	M10 ≤ M ≤ M20	0.20
	7.5 < fbk< 15	M10 ≤ M ≤ M20	0.15
	fbk ≤ 15	M10 ≤ M ≤ M20	0.10

The design resistances are obtained from the well-known relation:

$$f_d = \frac{f_k}{\gamma_m}$$

With γ_m equal to: $\gamma_m = 5$, AS control, (allowable stresses)
$\gamma_m = 3$, ULS control, (ultimate limit state)

Regarding deformation parameters, the Technical Standards consider the following elastic moduli for safety analysis and checks:

- Normal shear modulus of elasticity — $E = 1000\ fk$
- Tangential shear modulus of elasticity — $G = 0.4\ E$

CHAPTER 5 [3]

5.0 ULTIMATE LIMIT STATE CHECK.

The checks are made on the assumption that the plane sections remain plane, neglecting the tensile strength of the masonry. Each set of walls can be checked in the ultimate limit state or tension for the following loading conditions:
- Bending compression for lateral loads. (out-of-plane resistance and stability)
- Compression for in-plane actions
- Shear for in-plane actions.
- Concentrated loads.

In addition, checks must be carried out on the resistance of masonry beams and floor slabs, the collapse of which affects the response of the entire building as much as the collapse of the walls.

5.1 ECCENTRICITY OF LOADS

Conventionally, the actions on walls and on slabs are evaluated by simulating the masonry as a simple support for the slabs. In this case, the vertical loads acting on each "**pier**" are generally not applied to the center of gravity of the section, but each exhibits a certain eccentricity. To this is added the longitudinal eccentricity which is caused by the horizontal loads acting on the plane of the masonry.
These types of eccentricities are conventionally classified according to this scheme:

a- Total eccentricity of vertical loads e_s

b- Eccentricity caused by application tolerances e_a

c- Eccentricity caused by horizontal forces acting in a direction normal to the plane of the wall e_v

d- Eccentricity caused by horizontal forces acting in the plane of the wall e_b

5.1.1- Total ECCENTRICITY of vertical loads

This eccentricity is due to the vertical loads transmitted from the floor slabs that are directly supported on the "**wall**" being analyzed, as well as from the vertical loads coming from the masonry of the floor above. It depends on the geometric characteristics of the masonry and the distribution of loads brought about by the support of the slab on the walls: $\mathbf{e_s = e_{s1} + e_{s2}}$

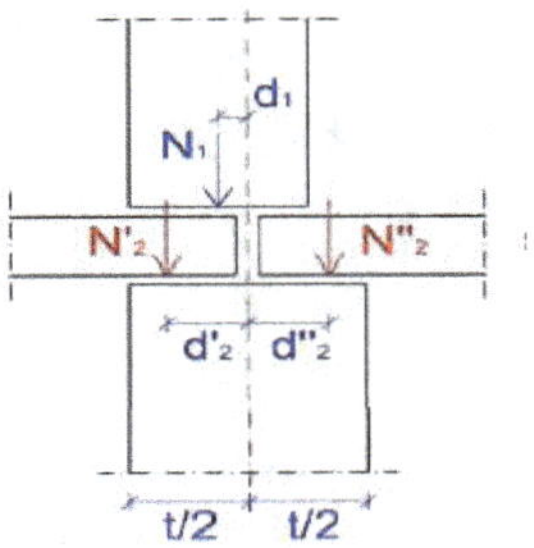

Fig. 5.1- Loads and eccentricities

where:
$\mathbf{e_{s1}}$ - refers to the eventual eccentricity position of the wall of the upper floor, referred to the average plane of the wall in the analysis.
$\mathbf{e_{s2}}$ - eccentricity of the support reaction of the floor slabs, above the section in the analysis.

$$e_{s1} = \frac{N_1 d_1}{N_1 + \sum N_2} \qquad e_{s2} = \frac{\sum N_2 d_2}{N_1 + \sum N_2} \tag{5.1}$$

where:
N1- is the load transmitted from the wall above
N2- is the reaction in the support of the slabs, on the wall in question .
d1- is the eccentricity of the load N1 in relation to the center of the wall to be verified.
d2- is the eccentricity of N2 in relation to the center of the wall to be verified.
These eccentricities are considered positive or negative based on the moment they give clockwise, or counterclockwise.

5.1.1- ECCENTRICITY due to implementation tolerance

The accidental eccentricity ea takes into account possible inaccuracies during the construction of masonry buildings and is determined by the fixed ratio:

$e_a = \frac{h}{200}$; where h – internal floor height.

5.1.1- **Eccentricity** due to horizontal forces acting in a direction normal to the plane of the wall ev

This eccentricity is calculated as: $\mathbf{e_v}=\frac{Mv}{N}$
$\mathbf{M_v}$- is the maximum bending moment, the consequence of actions perpendicular to the plane.
N- is the total normal force acting on the section we are verifying.

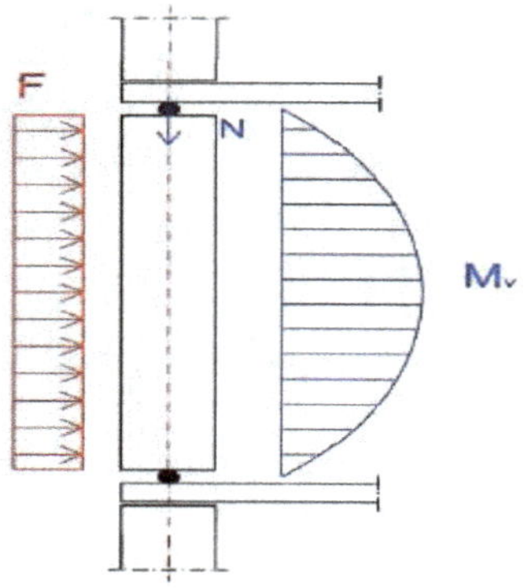

Fig. 5.2- Loads outside the wall plane

It is assumed that the wall has hinges at the floor levels. In the absence of openings, these hinges will also be in correspondence with the walls of the other direction, if they have an interaction of less than 6m. The eccentricities e_a, e_s, e_v are conventionally combined among themselves according to the following expressions:

$$e_1=e_s+e_a \qquad e_2=e_v+\frac{e1}{2} \tag{5.2}$$

The value $\mathbf{e_1}$ is adapted for the verification of walls in their extreme sections.
The value $\mathbf{e_2}$ is adapted for the verification of sections where the max. value of My is.
The eccentricity of the calculation cannot be accepted lower than ea. In any case the following must result:

$$\mathbf{e_1 \leq 0.33*t} \qquad \mathbf{e_2 \leq 0.33*t}$$

The eccentricity values were used to estimate the resistance reduction coefficients **Ø.**

5.1.4 ECCENTRICITY due to horizontal forces acting on the plane of the wall e_b.

The eccentricity caused by compression-bending in the plane of the wall is calculated:

$$e_b=\frac{Mb}{N} \tag{5.3}$$

where:
$\mathbf{M_b}$ - is the max. bending moment in the plane of the wall caused by horizontal actions acting on this floor.
N - is the total normal force acting on the section we are verifying.

5.2. PRESSURE - BENDING, from lateral loads (outside the plane of the wall)

- Reduction coefficient Ø.

A masonry element is under compression-bending action whenever the acting normal force N is not located at the center of the masonry section but exhibits an eccentricity.

$0 \leq x \leq \bar{h}$ (cracked section)

$\bar{h} \leq x \leq h$ (Uncracked section, totally reactive)

Ne zonen e paplasaritur kemi :

The presence of this eccentricity and the second-order effects lead to a reduction in the resistance of the panel. The masonry has a decrease in tensile strength so that the sections result in splits. Let us consider a wall (pier) loaded out of plane:

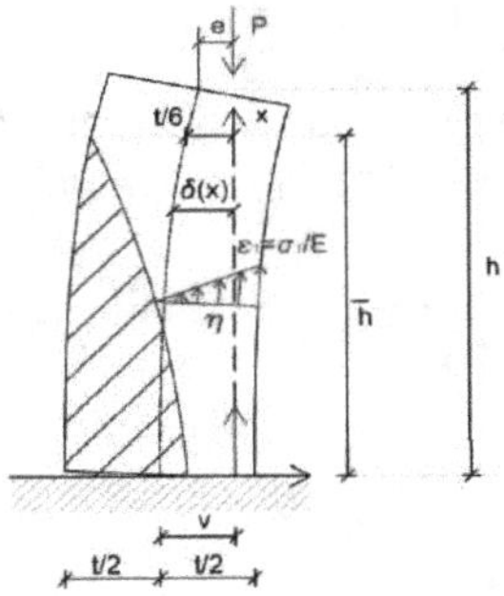

Fig. 5.3 Off-center load

$$EJ\delta'' = -P \cdot \delta$$

(5.4)

In the cracked area we have: $\delta'' = \frac{\varepsilon_1}{\eta}$ (curvature)

with $\varepsilon_1 = \frac{\sigma_1}{E}$ $\quad \sigma_1 = \frac{2P}{b \cdot \eta}$ $\quad \eta = 3\left(\frac{t}{2} - \delta\right)$

In this case the equation of the elastic line becomes: $EJ\delta'' + \frac{2Pt^3}{[27(t-2\delta)^2]} = 0$

(5.5)

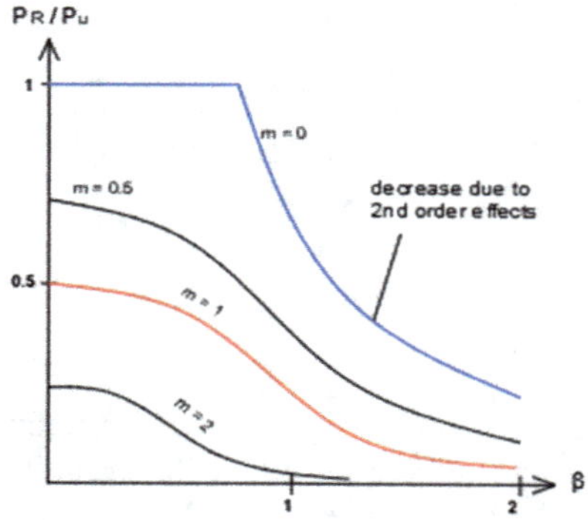

Assuming that the boundary conditions are:

δ(h) = e; δ(0) = v; δ'(0)= 0

can solve the problem caused by compression-bending for non-tensile materials. This problem was solved in 1971 by Sahlin, who proposed the above graph,

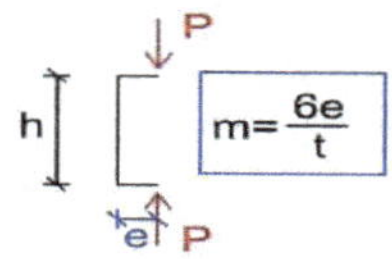

where m is:

$\beta = \frac{h}{b}\left(\sqrt{\sigma_u/E}\right)$ where σu – The ultimate tension in compression

If we note with:

$\boldsymbol{P_u = \sigma_u bt}$ - Ultimate bearing capacity of the element for concentrated load

$P_R = \phi P_u$ - ultimate bearing capacity of the element in compression-flexure.

$\boldsymbol{\phi = \phi(\lambda, m)}$ - reduction coefficient

$\lambda = \rho\frac{h}{t}$ - conventional wall flexibility, should be **λ≤ 20**,

ϱ - wall coupling coefficient.

$\frac{\ }{L} \geq 2$ ϱ = 0.45 ; $\frac{\ }{L} = 1$ ϱ = 0.71

$\frac{\ }{L} = \frac{2}{3}$ ϱ = 0.83 ; $\frac{\ }{L} \leq 0.5$ ϱ = 1.00

$\frac{\ }{L} \geq 4$ ϱ = 0.60 ; $\frac{\ }{L} = 2$ ϱ = 0.83

$\frac{\;}{L} = 1 \quad \varrho = 0.95$; $\qquad \frac{\;}{L} \leq 0.5 \quad \varrho = 1.00$

The masonry resistance reduction coefficient φ also depends on the conditions of the wall connection at its ends.

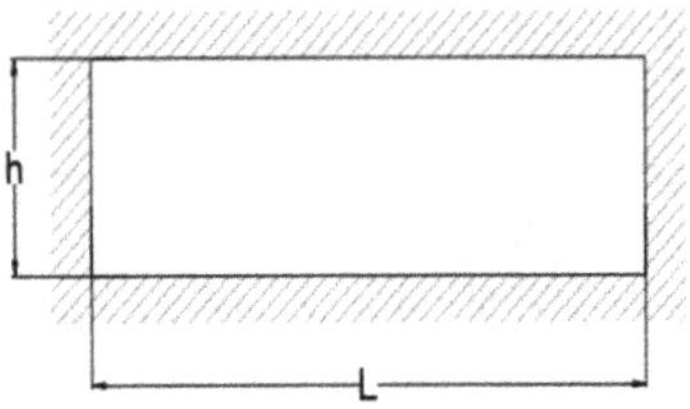

Tab. 5.1 The connection between flexibility h_0/t with the coefficient m= 6e/t

Flexibility ho/t	Coefficient of eccentricity m=6e/t				
	0	0.5	1.0	1.5	2
0	1.00	0.74	0.59	0.44	0.33
5	0.97	0.71	0.55	0.39	0.27
10	0.86	0.61	0.45	0.27	0.15
15	0.69	0.48	0.32	0.17	--
20	0.53	0.36	0.23	--	--

For values not included in the table, linear interpolation is permitted; extrapolations are not permitted under any circumstances. The tables and procedures are also valid for the effect caused by longitudinal eccentricity, by horizontal loads or in general by compression-bending in the plane of the wall.

Verification according to E.C. 2008

It consists of comparing the acting axial force and the resisting force, reduced to take eccentricity into account.

$$N_{Sd} \leq N_{Rd} = \Phi \cdot f_d \cdot A \qquad (5.6)$$

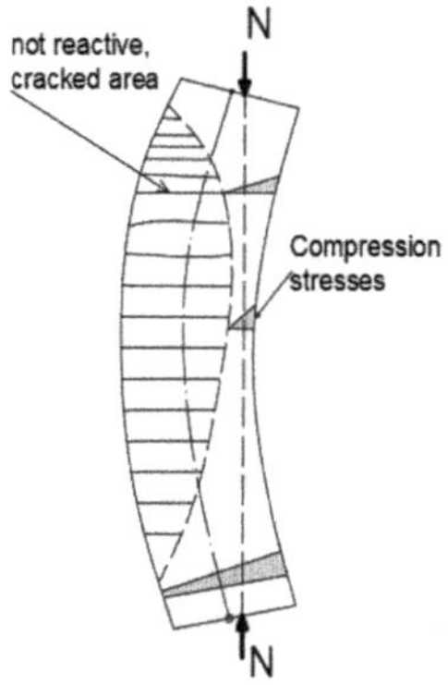

Fig. 5.4- Areas of tension and compression

Where:

N_{Sd} - is the design axial acting force
N_{Rd} – is the design axial resisting force
Φ - is the resistance reduction coefficient that takes into account the horizontal eccentricity of the loads and the wall deflection
f_d - is the design compressive strength
A - is the horizontal surface area of the wall in the section being verified.

5.3. PRESSURE WITH BEND for actions in the plane of the wall. [3]

Unlike previous standards, the verification of the bending compression in the plane of the wall is based on the assumption that the design resistance MRd in bending is greater than the bending moment MSd acting on the section being verified.
The state of failure from bending compression is related to the compression of the masonry by lateral compression. Moreover, if the axial load is small, we have large cracks from bending, and the masonry tends to develop a kinetic inversion similar to that of a solid body.
To simplify the expression of the resisting moment, we can use the concept of "stressed block", to express the compression diagram, in a similar way to what we usually do with reinforced concrete sections.

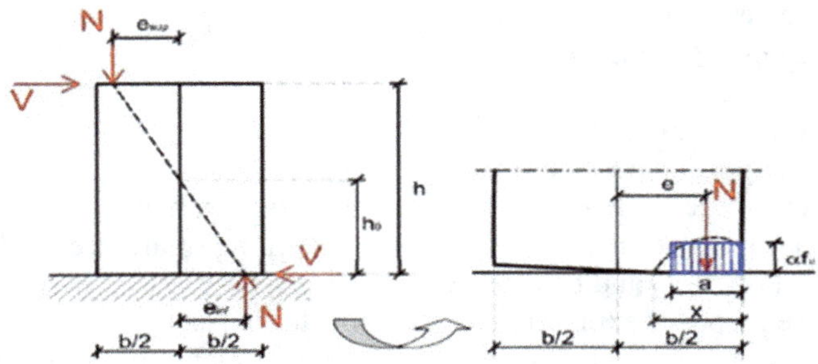

Fig. 5.5- The effect of eccentric compression and stress diagrams

Let's write the equilibrium equation:

N_{Sd} = **a*** f_d * a *t (vertical displacement equilibrium)
M_{Rd} = Nsd* (b –a) /2 (rotational equilibrium)

From the system of equations we have:

$$a = \frac{N_{Sd}}{\alpha \cdot t \cdot f_d} \qquad \sigma_o = \frac{N_{Sd}}{A} = \frac{N_{Sd}}{b \cdot t} \qquad a = 0.8x \tag{5.7}$$

$$M_{Rd} = N_{Sd}\left(\frac{b}{2} - \frac{1}{2}\frac{N_{Sd}}{\alpha \cdot t \cdot f}\right) = \frac{N_{Sd} \cdot b}{2}\left(1 - \frac{N_{Sd}}{\alpha \cdot b \cdot t \cdot f_d}\right) \tag{5.8}$$

$$M_{Rd} = \frac{\sigma_o \cdot b^2 \cdot t}{2}\left(1 - \frac{\sigma_o}{\alpha \cdot f_d}\right) \tag{5.9}$$

The corresponding value of the maximum shear is derived from the rotational equilibrium.:

$$V \cdot h_0 = N_{Sd} \cdot eRd_{inf}$$
(5.10)

$$V = \frac{\sigma_o \cdot b^2 \cdot t}{2 \cdot h_o}\left(1 - \frac{\sigma_o}{\alpha \cdot f_d}\right)$$
(5.11)

The 2008 regulations provide the following expression for this verification::

$$M_{Sd} \leq M_{Rd} = \frac{t \cdot b^2}{2}\frac{N_{Sd}}{A}\left(1 - \frac{N_{Sd}}{A\,\alpha\,f_d}\right) \tag{5.12}$$

Where:
M_{Sd} – is the design acting bending moment

M_{Rd} – is the design resisting moment
N_{Sd} - is the design acting axial force
t - is the wall thickness
b - is the total length of the wall (partition)
f_d - is the design compressive strength
A - is the cross-sectional area of the wall in the section being verified
α - is the coefficient that takes into account the filling of the stress diagram of the reactive section. It should be $\alpha \leq 0.85$.
Through the proposed expression it is possible to find an interaction zone between M_{Rd} and N_{Sd}. It is noted that the bending strength increases with the increase in the applied normal stress, then decreases again.

5.4. SHEAR from acting loads on the floor.

The failure modes of a mural panel depend on both its dimensions and the applied load and are classified into:
- Compression-flexural failure VRd,p (when the value of $h/b > 2$ - thin elements)
- Sliding failure VRd,s (occurs for low values of axial force)
- Diagonal cracking failure VRd,f (occurs for values of $h/b < 1.5$ - short elements)

The shear resistance is equal to: VRd = min (VRd,p; VRd,s; VRd,f)

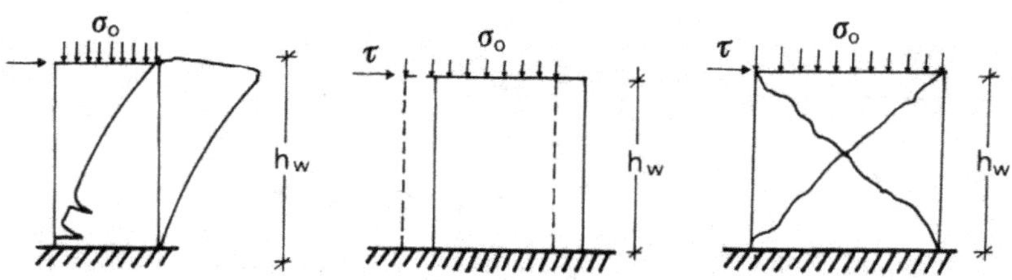

a) failure from b) sliding failure
c) failure from
compression-flexure
diagonal cracking

5.4.1. DESTRUCTION from bending compression.

In the case of failure from compression with bending, the value of the shear resistance VR_d is determined by the compression conditions of the masonry pressed into the lower base of the panel, as already seen in §4.7.6.

$$V_{Rd,p} \cdot h_o = N_{Sd} \cdot eRd_{inf} \tag{5.13}$$

$$V_{Rd,p} = \frac{N_{Sd} \cdot b}{2h_o}\left(1 - \frac{N_{Sd}}{\alpha \cdot b \cdot t \cdot f_d}\right) \tag{5.14}$$

where :
N_{sd}- is the design axial force
t - is the wall thickness
b - is the overall length of the wall
f_d - is the design compressive strength
h_0 - is the height of the point where the moment becomes zero
α - is the coefficient that takes into account the filling of the stress diagram of the reactive section. It must be $\alpha \leq 0.85$

5.4.2.– DESTRUCTION from slide.

In the case where the vertical load is small, failure by sliding at the mortar joints may occur, assuming a "Columbian" behavior of the material.

$$f_{vk} = f_{vk0} + 0.4\,\sigma_0$$

The latest regulations refer to this model, evaluating the resistance of the masonry as the product of the unit resistance of the wall's reactive surface (compressed area).

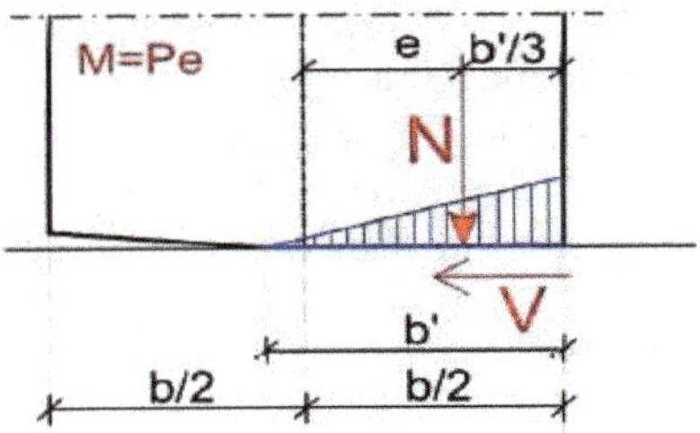

$$b' = \left(\frac{3}{2} - \frac{3e}{2}\right) b \qquad e = \frac{V \cdot h_o}{P} \qquad \sigma_0 = \frac{P}{b \cdot t} \tag{5.15}$$

$$V = f_{vd} b' \cdot t = \left[\left(\frac{3}{2} - \frac{3Vh_0}{Pb}\right) f_{vd0} + 0.4\sigma_0\right] \frac{b \cdot t}{\gamma_m} \tag{5.16}$$

$$V_{Rd,s} = \frac{1.5 f_{vk0} + 0.4\sigma_0}{1 + \frac{3h_0}{b\sigma_0}\frac{f_{vk0}}{\gamma_m}} \frac{b \cdot t}{\gamma_m} \tag{5.17}$$

The 2008 regulations provide for this case to be verified according to the following expression:

$$V_{sd} \leq V_{Rd,s} = \beta \cdot A \cdot f_{vd} \tag{5.18}$$

Where:

V_{sd} – is the design shear force
$V_{Rd,s}$ - is the design shear force in sliding
A – is the cross-sectional area of the wall for verification
f_{vd} – is the design shear resistance
β - is the coefficient of partition into parts of the wall created by the external moment Msd, regardless of the eccentricity $e_b = M_{sd} / N_{sd}$.

In accordance with EC 1987, it is valid:

$$\beta = \{ 1 \text{ nqs } \frac{6eb}{b} \leq 1\} \qquad \beta = \{ (\frac{3}{2} - \frac{3eb}{b}) \text{ nqs } 1 < \frac{6eb}{b} \leq 1.3 \}$$

5.4.3. **DESTRUCTION** from diagonal cracks

It is assumed that shear failure occurs when the principal (macroscopic) tensile stress reaches the limit value, accepting as the conventional tensile strength of masonry the relation proposed by Turnsek and Cacovic (1971).

$$V_{Rd,f} = b \cdot t \frac{1.5\tau_{0d}}{\xi} \sqrt{1 + \frac{\sigma_0}{1.5\tau_{0d}}} = b \cdot t \frac{f_{td}}{\xi} \sqrt{1 + \frac{\sigma_0}{f_{td}}}$$

(5.19)

where:
$V_{Rd,f}$ - is the design shear strength resistant to diagonal cracking
f_{td} - is the design tensile strength for diagonal cracking ($f_{td} = 1.5\, \tau_{od}$);
τ_{od} - is the corresponding reference shear strength of the masonry, equal to f_{vd0}
ξ - is the correction coefficient related to the stress distribution in the section, depending on the $\frac{}{b}$ ratio of the wall.

$$\xi = \begin{cases} 1.0 & \text{nqs } \frac{}{b} \leq 1.0 \\ \frac{}{b} & \text{nqs } 1.0 \leq \frac{}{b} \leq 1.5 \\ 1.5 & \text{nqs } \frac{}{b} \geq 1.5 \end{cases}$$

(5.20)

5.4.4 ZONA τ-σ

Shear resistance depends on:
- Slide mechanism per $\sigma \leq \sigma_1$
- Diagonal cracking mechanism per $\sigma_1 \leq \sigma \leq \sigma_2$
- Destruction mechanism per $\sigma > \sigma_2$

In fact, diagonal cracking according to the Turnsek-Cacovic mechanism only occurs when the masonry texture is irregular (existing stone buildings).

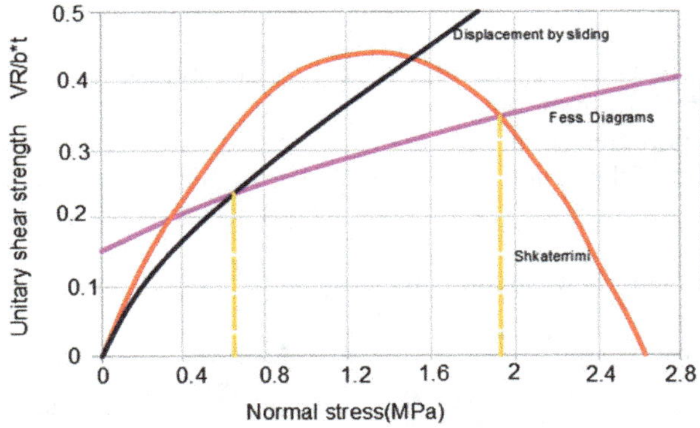

f (MPa)	6.2
fvk0 (MPa)	0.3
γm	2.0
ξ	1.5

Fig. 5.7- Dependence of the $V_{Rd}/b*t$ ratio on normal stresses

5.5. CONCENTRATED loads [3]

It is a test that was not provided for by previous regulations, and consists of checking that the design value of the concentrated load is less than the corresponding resistance:

$$N_{sdc} \leq N_{Rdc} = \beta_c \cdot A_c \cdot f_d \quad (5.21)$$

where:

N_{Sdc} – is the design concentrated load
N_{Rdc} - is the design resistance to concentrated loads
A_c - is the support surface
f_d - is the design resistance in compression
β_c - is the amplification coefficient for concentrated loads, evaluated as a function of the type of masonry, as shown below:
For class 1 resistant elements, according to the 2005 norms, we have:

$$\beta_c = \left(1 + 0.3\frac{a_1}{h_c}\right)\left(1.5 - 1.1\frac{A_c}{A_{eff}}\right) \quad \text{ne te cilen} \quad (5.22)$$

$$1 \leq \beta_c \leq \begin{cases} 1.5 \\ 1.25 + \dfrac{a1}{2hc} \end{cases}$$

Where:
a_1 - is the minimum distance between the edge of the support and the wall
h_c - is the height of the wall from the support level
A_{ef} - is the effective support area equal to l_{ef}*t
l_{ef} - is the effective length of the support, calculated as in the figure and in any case lef $\leq$ 2.2 Ac / t
t - is the wall thickness
If the elements are not of class 1, we assume $\beta = 1.0$

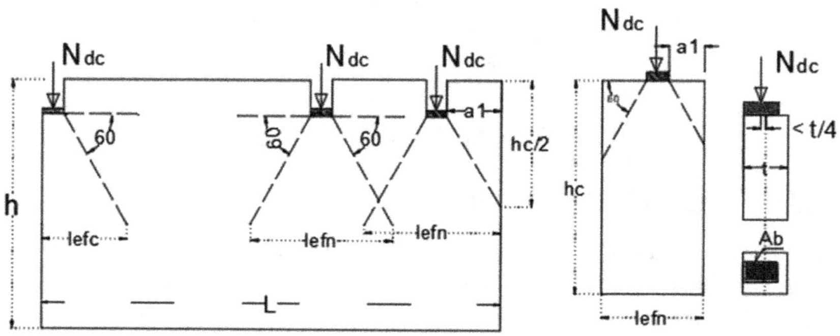

Fig. 5.8- Angle of distribution of concentrated force in masonry

CHAPTER 6 [3]

(MASONRY CONTROLS)

6.1 CONTROL OF A FLOOR BELT BEAMS IN THE MASONRY.

The structural function of the horizontal subfloor bands (spandrels) is not secondary, as they provide a significant degree of cooperation between the vertical bands (piers) and influence the overall behavior of the multi-story wall, in the load-bearing masonry. The stress state created by the action of horizontal forces on the spandrels is analogous to that shown in the figure below. The mechanical behavior of a spandrel can be studied in analogy to that of a pier, taking into account some important differences:

- In the case of regular masonry (e. g. square, regular block bricks masonry), the orientation of the mortar layers is parallel to the axis of the element, which is horizontal.

- The axial action (resulting from compression parallel to the horizontal axis), caused only by the gravity load, is usually low, or zero.

The cooperation that may be transmitted by the subfloor bands is mainly a function of the compression by which they are excited in the horizontal direction. Only this compression actually supplies the "bending" resistance that prevents the activation of the overturning mechanism shown in the figure.

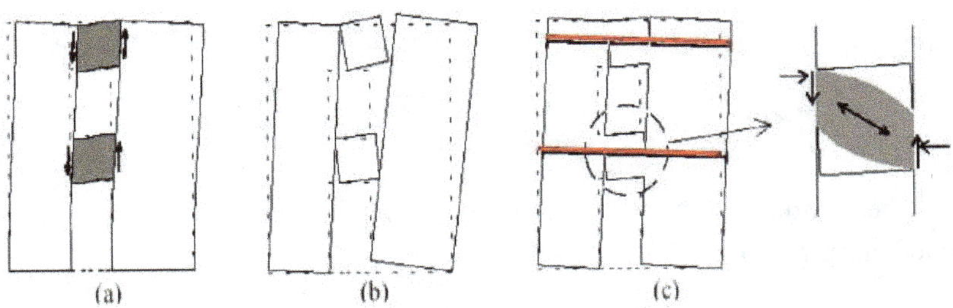

Fig. 6.1- Spandrel-pier interaction

In the absence of normal forces, beams tend to rotate rigidly and push away the vertical members, completely overturning the static system. In fact, in the absence of compression, the resisting moment and the resisting shear together become zero.

If P=0

$$\sigma_0 = \frac{P}{b \cdot t} = 0$$

$$M_{Rd} = \frac{\sigma_0 \cdot b^2 \cdot t}{2}\left(1 - \frac{\sigma_0}{\alpha \cdot f_d}\right) = 0$$

$$V_{Rd} = \frac{\sigma_0 \cdot b^2 \cdot t}{2 \cdot h_0}\left(1 - \frac{\sigma_0}{\alpha \cdot f_d}\right) = 0 \qquad (6.1)$$

The presence of horizontal ties that resist the torsion of the beams is essential. The load-bearing capacity of the beams against bending and shear is closely related to the dimensions of the horizontal connections (connecting bands, tie rods, etc.). The presence of ties that resist the rupture of the pier generates an increase in compression in the connecting elements, which increases the resistance to shear-flexure: thus, a mechanism with inclined columns is created, similar to that of the bending of vertical piers under the action of compression-flexure.

6.2 -CONTROL in compression-bending

a) If, through the analysis of the structure, the characteristics of the actions on the element in terms of N, M, Q result to be large, the procedure is similar to what we saw for the walls.

b) In the case where the axial action is not large, but there are horizontal elements capable of resisting traction, that is, which cause a compression in the wall (tie rods or belts, etc.), the resistance to compression-flexure can be determined according to:

$$M_{Rd} = \frac{H_p \cdot}{2}\left(1 \frac{H_p}{0.85 \cdot f_d \cdot h \cdot t}\right) \qquad (6.4)$$

H_P - is the minimum between the tensile strength of the horizontally placed tensioned element and the value $0.4 * f_{hd} * h * t$

f_d - is the design compression resistance of the masonry in the horizontal direction;

h - is the height of the floor joist.

t - is the thickness of the floor band.

a) Alternatively, we can make the following reasoning:
The minimum axial force P_{min} is determined, which must be guaranteed by the horizontal braces, so that the horizontal element is able to resist the design moment M_{Sd}.

$$\begin{cases} M_{Sd} = \alpha * f_{hd} * a * t * \left(\frac{\ }{2} - \frac{a}{2}\right) \\ P_{min} = \alpha * f_{hd} * a * t \longrightarrow a = \frac{Pmin}{\alpha * fhd * t} \\ \text{me } \alpha = 0.85 \text{ dhe } a = 0.8x \end{cases} \quad (6.5)$$

Let's replace α to M_{sd}:

$$M_{Sd} = \frac{P\ min}{\underline{h}\ 2} * \left(h - \frac{P_{min}}{\alpha \cdot f_{hd} \cdot t}\right) \quad (6.6)$$

$$Pmin = \frac{\alpha \cdot h_{hd} \cdot h \cdot t}{2} \left(1 - \sqrt{1 -}\right) \frac{\overline{\quad}}{\alpha \cdot h_{hd} \cdot h^2 \cdot t} \quad 8M_{Sd} \quad (6.7)$$

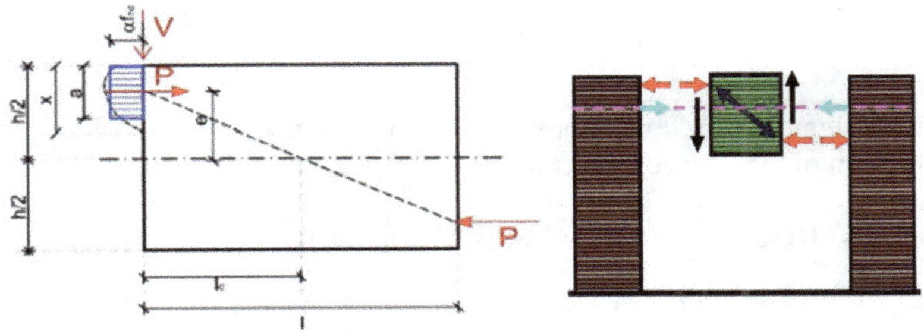

6.3 SHEARING CONTROL.

The value of the shear force in ordinary masonry beams will be equal to the minimum value of the following 4 mechanisms:

a- Pure shear failure $V_{Rd,t}$
b- Compression failure with bending $V_{Rd,p}$
c- Sliding failure $V_{Rd,s}$
d- Diagonal crack failure $V_{Rd,f}$

$$V_{Rd} = \min(V_{Rd,i})$$

6.4 The DESTRUCTION in a clean shear.

The shear resistance VRd,t of cooperating beams in ordinary masonry, in the presence of a floor band, or of a bending-resistant lintel, with effective fixation at the ends, can be calculated in a simplified manner as:

$$V_{Rd,t} = h \cdot t \cdot f_{vd0}$$

(6.8)

where:
f_{vd0} – is the design shear resistance in the absence of compression.
h - is the height of the floor slab.
t - is the thickness of the floor slab.

6.5 The DESTRUCTION in bending pressure. [3]

The shear resistance $V_{Rd,p}$ of connection beams in conventional masonry with respect to flexural collapse can be calculated according to:

$$V_{Rd,t} = \frac{2M_{Rd}}{l} \quad \text{with} \quad M_{Rd} = \frac{H_p \cdot h}{2}\left(1 - \frac{H_p}{0.85 \cdot f_d \cdot h \cdot t}\right) \quad (6.9)$$

where:
M_{Rd} – is the design moment for compression-flexure resistance.
l - is the span of the connection beams

6.6 - The DESTRUCTION in slide (according to Coulomb)

In an analogous way on the full vertical pier with N = P_{min}

$$V_{Rd,s} = \frac{1.5 f_{vk0} + 0.4\sigma_0}{1 + \frac{3l_0}{h\sigma_0}\frac{f_{vk0}}{\gamma_m}}\frac{h \cdot t}{\gamma_m} \quad \sigma_0 = \frac{P_{min}}{h \cdot t} \quad (6.10)$$

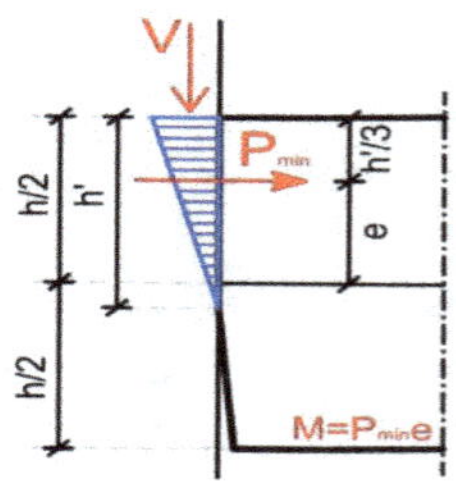

6.7- DESTRUCTION from diagonal cracks

Destruction in the shear is assumed according to the report proposed by Turnsek, Cacovic. (1971):

$$V_{Rd,f} = h \cdot t \frac{1.5\tau_{0d}}{\xi} \sqrt{1 + \frac{\sigma_0}{1.5\tau_{0d}}} = h \cdot t \frac{f_{td}}{\xi} \sqrt{1 + \frac{\sigma_0}{f_{td}}} \qquad \sigma_0 = \frac{Pmin}{h*t}$$

(6.11)

where:

$\mathbf{V_{Rd,f}}$ – is the design shear strength from diagonal cracks.
$\mathbf{f_{td}}$ - is the design resistance, tensile strength for vertical cracks (ftd = 1.5 τ_{0d})
τ_{0d} – is the corresponding reference shear strength of the masonry equal to $\mathbf{f_{vd0}}$.
ξ– is the correction factor, which is related to the distribution of stresses in the section; it depends on the wall ductility. It can be taken:

$$\xi = \begin{cases} 1.0 \text{ if } \frac{l}{} \leq 1.0 \\ \frac{l}{} \quad \text{if } 1.0 < \frac{l}{} < 1.5 \\ 1.5 \text{ nqs } \frac{l}{} \geq 1.5 \end{cases} \qquad (6.12)$$

The four mechanisms can be presented in a single diagram as a function of axial load P_{min}

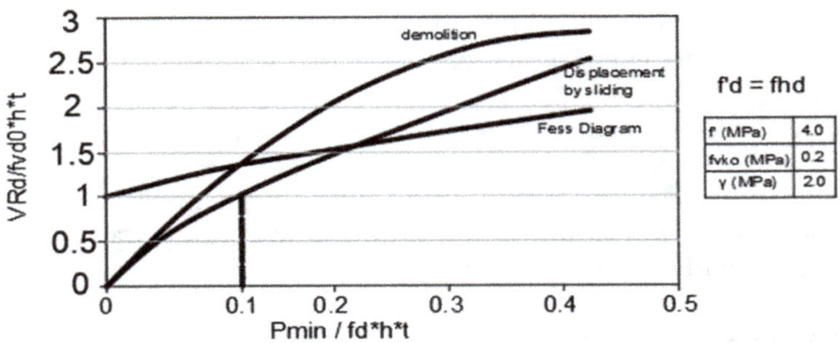

Fig. 6.3 - Shear – Axial load interaction diagram

6.8- LIMIT STATES OPERATION **CONTROL**

The limit state checks of the wall sets may not be carried out except in the following cases:

- The need to limit the expansion of cracks. The check is accompanied by the almost permanent combination by verifying that, with the hypothesis of zero tensile resistance and linear stress distribution, the cracks are limited to the thickness of the plaster.
- For buildings with a number of floors greater than 4, the displacement $\mathbf{d_r}$ between floors must be checked, which in frequent combinations must respect the following limit, where h is the height of the mezzanine:

$$d_r \leq 0.003 \cdot h \tag{6.13}$$

6.8.1 **Permissible tensions** control

For simple buildings, the use of the allowable stress control method is foreseen, through a simplified dimensioning provided that certain restrictions are respected. Previous regulations provided for:
- Compression with bending control
- Shear in floor actions control
- Concentrated loads (previously unforeseen) control
The stresses are calculated in the combination of particular loads of the limited action state, considering $\psi_i = 1$.

6.8.2 Pressure with bending control

The control is satisfied if:

$$\sigma = \frac{N_d}{\Phi_1 \cdot \Phi_t \cdot A} \leq f_d \tag{6.14}$$

where:
Nd- is the design axial force
A – is the cross-sectional area of the wall in the verification section
f_d - is the design compressive strength, estimated by the tension verification method
Φ_t – is the reduction coefficient for transverse eccentricity, analogous to that illustrated in § 4.7.5;
Φ_l – is the reduction coefficient for longitudinal eccentricity calculated for an eccentricity equal to $e_l = M_d / N_d$. It is calculated in the same way as illustrated in § 4.7.5, with $M = 6e_l/t$ e and: $h_0 / t = 0$.

6.8.3- Shear control for actions in the wall plane

The control is satisfied if:

$$\tau = \frac{V_d}{\beta \cdot A} \leq f_d \tag{6.15}$$

where:
Vd- is the design shear force

A – is the cross-sectional area of the wall (partition) in the verification section
fd – is the design shear resistance, estimated by the tension control method.
β - is the coefficient of fragmentation of the wall section, created by the external moment Md,
which depends on the eccentricity: e_b = Md / Nd and is given by the relations:

$$\beta = \begin{cases} 1 & nese \quad \frac{6eb}{b} \leq 1 \\ \frac{3}{2} - \frac{3eb}{b} & nese \quad 1 < \frac{6eb}{b} \leq 1.3 \end{cases} \qquad (6.16)$$

6.8.4 – Concentrated loads control

The control is satisfied if: $$\sigma = \frac{N_{cd}}{\beta_c \cdot A_c} \leq f_d \qquad (6.17)$$

where:
N_{cd} – is the design concentrated force
A_c – is the support surface
f_d – is the design compressive strength, calculated according to the tension control method.
β_c - is the amplification coefficient for concentrated loads, evaluated according to the type of masonry, as shown in § 4.7.8:

6.9 - **Normative requirements**
6.9.1- Concrete belts in masonry

The rules always require the need to conceive the building with load-bearing walls as a three-dimensional box-shaped structure. For this purpose, they must ensure the appropriate connections between the walls and the floor slab, as well as the walls between them.
The latest EC provisions do not explicitly mention the characteristics of the beams as illustrated in the previous decrees, in which, in the paragraph of masonry buildings, in seismic areas, this argument is developed in detail and with more limitations.
For structures in non-seismic areas, we can refer to the EC 1987 provisions, summarized below:

"In floor slabs, the belts (girdles) will generally be made of reinforced concrete, with a width equal to at least 2/3 of the wall below them, and in any case not less than 12 cm and a height at least equal to that of the slab, and in any case not less than half the thickness of the wall. The minimum reinforcement of the beams must be at least 6 cm2, with a diameter of not less than 12 mm. The rods must be 6 mm in diameter and placed at a distance not exceeding 30 cm. The minimum requirements determined in the non-seismic zone are best summarized in the diagram below:

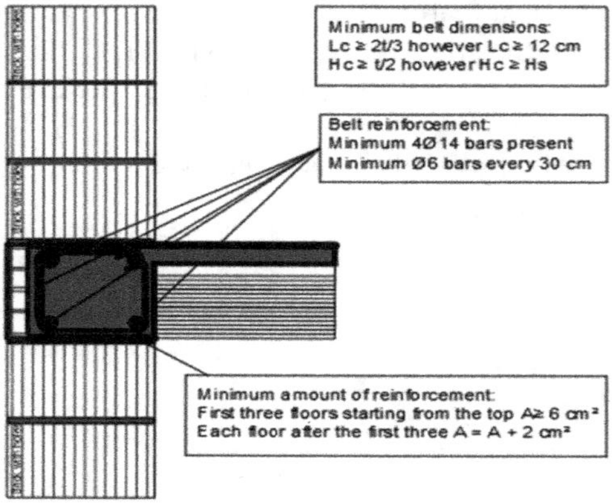

Fig. 6.4 Belts in masonry

6.9.2- Minimum wall thickness

EC 2008, in analogy to the previous EC, indicates the minimum wall thicknesses depending on the type of resistant element used:

- Muratura ne element rezistent artificiale te plotë — 120 mm
- Masonry in artificial resistant element half-full — 200 mm
- Masonry in resistant element with holes — 250 mm
- Squared stone masonry — 240 mm
- Listed masonry — 400 mm
- Non-squared stone masonry — 500 mm

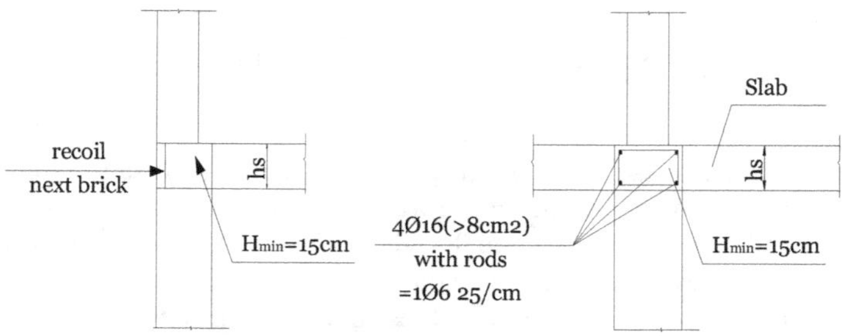

Fig. 6.5- Details of masonry belts

CHAPTER 7- SEISMIC VULNERABILITY IN EXISTING BUILDINGS [10]

7.1 - DEFINITION OF THE VULNERABILITY FUNCTION

In general, a vulnerability function is a relationship that defines the expected damage to a building, or a class of buildings, as a function of ground motion (Figure 3.1). The two key elements of a vulnerability analysis are the building capacity and the seismic demand. To estimate the damage **D**, the building's ability to resist (building capacity) must be compared with the stresses on the structure due to the ground motion from the earthquake (seismic demand).

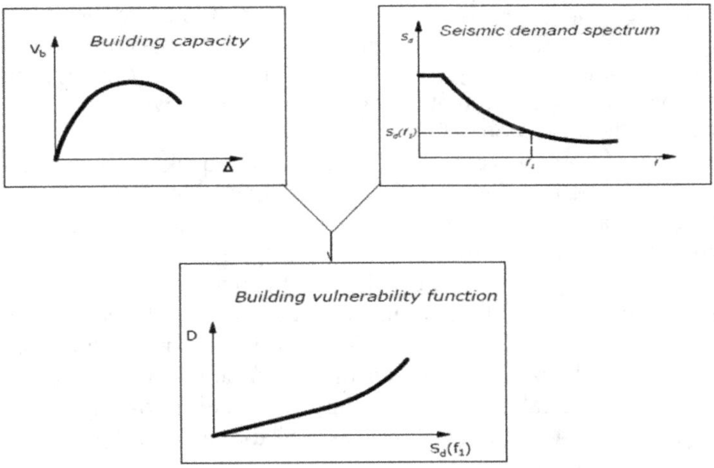

Fig. 7-1 Principles of Vulnerability Function

In earthquake engineering, the capacity of a building to resist seismic action is represented by a capacity curve, which is defined as the base shear force Vb acting on the building as a function of the horizontal displacement Δ at the top of the building, often also referred to as a pushover curve. The shear capacity of a building refers to the maximum base shear force V_b that the building can withstand, and the displacement capacity refers to the ultimate displacement at the top of the building Δ_b. Similarly, the shear capacity of any structure or structural element refers to the maximum shear force it can withstand, and the displacement capacity refers to its ultimate horizontal displacement. To express seismic demand, until recently, "intensity" was used almost exclusively.
This is a descriptive parameter of an earthquake based on observations of the effects of the earthquake on the environment. It has the advantage until the historical earthquake data are available. However, frequently, the information on

the true motion dispersed and the empirical relationships between intensity and peak ground acceleration vary greatly. Some methods use peak ground acceleration as the parameter defining the earthquake. However, in this case, not only information on the duration of the earthquake is lost, but also information on the frequency composition.
Thus, a better parameter is the spectral acceleration, or as we will see spectral displacement
A method to assess the vulnerability of existing buildings for the purpose of comparison with other methods or with test results, the peak ground acceleration **Sa** has also been used.

The shaking of the ground due to an earthquake does not occur in a fixed direction; on the contrary, in a horizontal plane, the direction of the shaking of the ground varies, including all angles from 0 to 360°. However, the largest amplitudes of the shaking of the ground usually occur in one direction, while normal to the direction of the largest amplitude, they are much smaller. Thus, the stresses in the building are dominant in the direction of the largest amplitudes, which is referred to in the following as the "direction" of the earthquake.
For regular buildings, it is common practice in earthquake engineering to consider the earthquake action (i.e. the direction of the greatest amplitude of ground shaking) separately, in two orthogonal directions, usually corresponding to the principal axes of the building, using plane analyses.
Thus, for a building, two vulnerability functions are calculated. For earthquake scenarios, the direction of an earthquake is usually not taken into account and, based on the two vulnerability functions, in the two main directions, a single representative vulnerability function of the building must be calculated. This representative vulnerability function must describe the overall behavior of the building, i.e. it must be a kind of "expression" of the two vulnerability functions, in the two main directions.
Choosing the more unfavorable vulnerability function of the two will lead to a "worst case scenario", which is not desirable in the case of earthquake scenarios, as it can be assumed that on average, the building "behaves better". For very irregular buildings, the dual vulnerability function, in the two main directions, can be very different and thus the direction of the earthquake action plays an important role. Since this is not taken into account, the inaccuracy resulting from the introduction of a single representative function increases.

7.2 - **BUILDING CAPACITY CURVE** [10]

7.2.1 - General considerations

The capacity curve is generally constructed to represent the first mode of response of a building, based on the assumption that the structure responds to a seismic input mainly in its fundamental mode of vibration. (Figure 7.2). Thus the horizontal force distribution over the height of the building should conform to the first modal form, the lateral forces at the floors being proportional to the product of the mass and the fundamental modal form:

$$F_i = \frac{m_i \phi_i}{\sum m_i \phi_i} \cdot V_b . \tag{7.1}$$

- mi is the concentrated mass and **φi** is the displacement in the first mode at the level of the **i-th** floor.
- Vb is the shear force at the base and corresponds to the sum of the lateral forces of the floors.
However, the basic modal form is usually not known initially.
To determine it, it is assumed that:
- depending on the structural system, walls or frames, the first modal shape corresponds to a parabola (bending displacements).
- or to an almost straight line (shear displacement),
- or lateral forces in accordance with the code can be applied at each storey and the resulting displacements are used as an approximation of the basic first modal shape:

$$F_i = \frac{m_i h_i}{\sum m_i h_i} \cdot V_b . \tag{7.2}$$

hi - is the height of the i-th floor.

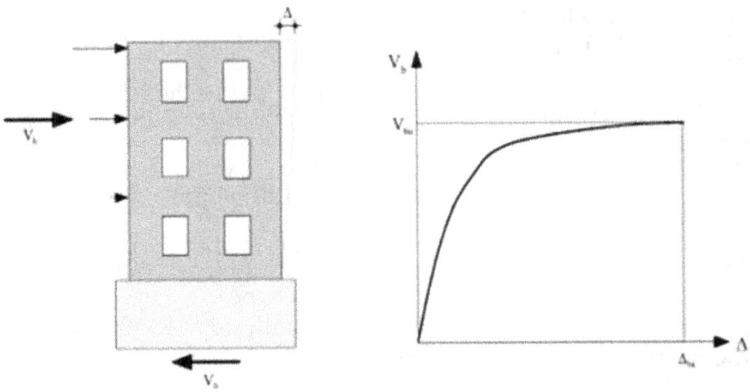

Fig. 7.2- Capacity curve

Since the results are not very sensitive to the distribution of forces, any reasonable distribution can be made. By increasing the total horizontal force acting on the building, or displacements respectively, the capacity curve of the building is obtained, by plotting the shear force at the base Vb as a function of the displacement D at the top. (Fig. 7.2).
As the application of horizontal forces, under the first action, considered monotonic, the effects of cyclic loading must be included in a further step (see Section 4.2.5 for masonry buildings).

7.2.2. – Identification of structural and non-structural elements.

In every building, structural and non-structural elements must be distinguished.

In contrast to the design of a building, where the structural system is chosen and therefore known, the assessment of the building requires first the identification of the structural system, with all its elements, since only these contribute to the capacity of the building.
Non-structural elements only add weight. In the case of concrete buildings, the identification of structural elements is usually done without difficulty, as the structural walls and/or frames can be easily identified. In the case of masonry buildings, this is usually less clear, because all walls, (both facade and interior walls) are made of masonry and there is often no clear distinction between them. However, it is common practice to consider all walls with a thickness of $t \geq 12$ as structural walls, i.e. those that act to support vertical and horizontal forces. Not all walls with a thickness of $t \geq 12$ need to be structural, especially in modern buildings there are more and more non-structural walls with a thickness of $t \geq 12$, however, in the absence of more detailed information the above assumption seems reasonable.

7.2.3 – Terminology and structural models

Considering the building in Figure 7.3, regardless of the material (masonry or concrete) the terminology used in the construction of the capacity curve of a building is given below:

1. A wall is defined as a structural element of the building with length l_W and a height equal to the total height of the building H_{tot}.
2. A wall element can be any part of it with length l_W and height h (not shown in Figure 7.3).
3. A pier is a wall element, with a length and height equal to the height of the opening bordering it, which may be a window or a door (shown by the lightly shaded areas).
4. Spandrels are those parts of the building which extend between two openings in the vertical direction, joining the walls in one plane (shown by the heavily shaded areas).
5. All walls in one plane connected to the floors and spandrels constitute a plane wall. Thus the facade of a building constitutes a plane wall, as do all walls in one plane, inside the building.

The wall panel is defined as a part of a plane wall, with any length and height equal to the height of the floor (not shown in Fig. 7.3).
Also shown in Fig. 7.3 are the applied forces:

1. Equivalent horizontal earthquake forces are assumed to be applied at the floor level where the mass is highest.
2. Vertical loads include the self-weight of the structure, as a volume force, the overloads (non-structural elements) and temporary loads, which are applied at the interfloor levels.

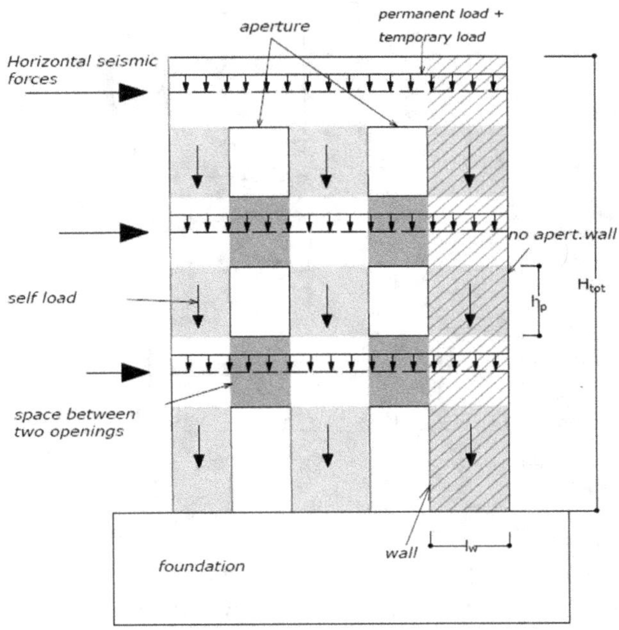

Fig. 7.3 – Masonry element designations

Due to the fact that the walls are connected by the spandrels, a coupling effect is created. Depending on the spandrel section, this coupling effect will be greater or less. In the absence of spandrels, where the walls are connected only by the slabs (usually for concrete buildings) the coupling effect is negligible and the walls can be considered as cantilever. For deep spandrels (often found in masonry buildings) the coupling effect is considerable and must be taken into account in the calculations.

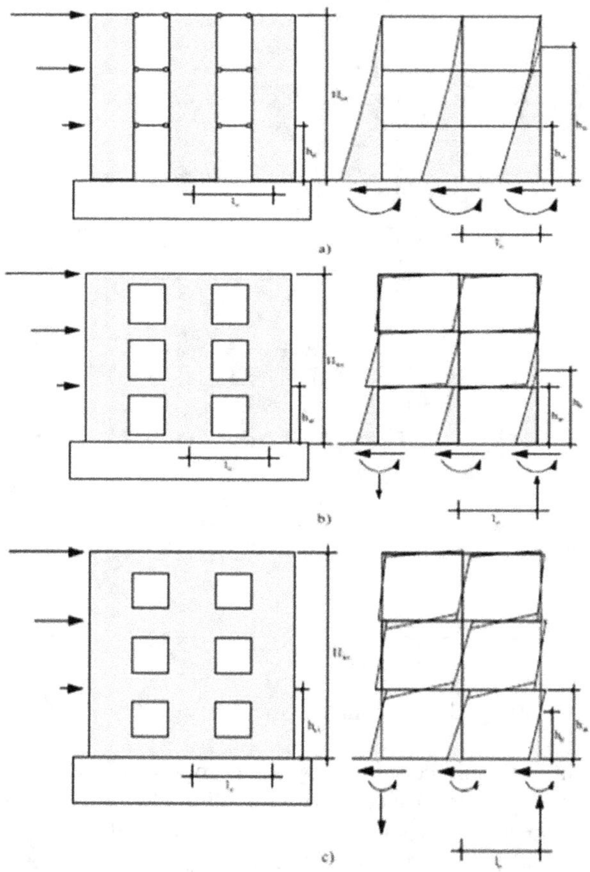

Fig. 7.4- *Bending moment distribution for three cases of coupled walls:*
a) negligible effect of connection (cantilever walls),
b) intermediate coupling effect and
c) strong effect of connection, due to horizontal earthquake forces and corresponding reactions.

The connected wall system can be analyzed using a frame model. In general, any plane wall can be considered as a connected wall system, the case of cantilever walls being a "limiting case", where the stiffness of the spandrels becomes negligible with respect to the stiffness of the walls and consequently the effect of the connection is reduced to zero.

Figure 7.4 shows the bending moment distribution for three cases of connected walls, under the action of horizontal forces. Figure 7.4-a) shows the case where the walls are connected only by the floors and thus the connection effect is negligible, the whole system can be viewed as cantilever walls. Figure 7.4 c) shows the case of very deep spandrels giving a considerable connection effect and Figure 7.4 b) shows an intermediate case, with a small connection effect.

In the case of cantilevered walls (Figure 7.4 a), the total overturning moment, due to the applied horizontal forces, is resisted by the walls alone, in proportion to their stiffness, resulting in very high bending moments at the base of the walls.
In the case of strongly coupled walls (Figure 7.4 c), the total overturning moment, due to the applied horizontal forces, is resisted mainly by the high normal forces in the outer walls, resulting from the vertical shear forces transmitted by the spandrels.
The bending moments at the base of the walls are therefore quite small compared to those of a cantilever wall. In the intermediate case (Figure 7.4 b) the frame action is smaller and so a portion of the total overturning moment resisted by the walls is increased, while the normal forces are reduced. For regular frames, the extent of the coupling effect can be expressed by a single parameter, the height h_0 of the moment being zero (Figure 7.4). The smaller the value of h_0, the greater the coupling effect. For infinitely stiff spandrels, the limiting value of $h_0 = 0.5*h_{st}$. As the coupling effect decreases, the height h_0 of the zero moment increases, eventually becoming greater than h_{st}.
The value of h_0 depends on the ratio of the bending stiffness of the spandrel to the bending stiffness of the pier (EIsp/lo)/(EIp /hst). Figure 3.6 a) gives the ratio of h_0/h_{st} as a function of (EIsp /lo)/(EIp /hst) for a 3x3 frame (where the first number indicates the number of walls and the second number the number of stories), as shown in Figure 7.4, with a uniform story height hst, a uniform distance between the centerlines of the walls lo , and a horizontal triangular force distribution. For very deep spandrels, (Eisp/lo)/(EIp /hst), the ratio of h0/hst tends to 0.6, for the outer piers and 0.5 for the inner ones. For very flexible spandrels, h0 can reach hst, and in the limit state, when the stiffness of the spandrel is reduced to zero, the ratio of h0/hst will be greater than one, the bending moment distribution of the structure approaching the bending moment distribution of cantilever walls.
Note that for h_0/h_{st}, h_0 does not indicate the height of a true zero moment point, but corresponds to the height of the extrapolated zero moment of the pier. The transfer of horizontal inertia forces of the floors onto the walls must be ensured by floor-to-wall connections. In the case of concrete floors, the connection between the floor and the walls is usually good, and thus the transfer of horizontal forces onto the walls can be guaranteed.
In the case of wooden floors, the connection between the subfloor and the walls can be very weak, if not improved with special interventions such as steel bar anchorages, and the transfer of forces to the walls cannot be guaranteed, leading

to an uneven distribution of forces, overstressing some walls, while others remain almost unreinforced. (This is not analyzed below).

7.2.4 The realisation OF CAPACITY CURVES [10]

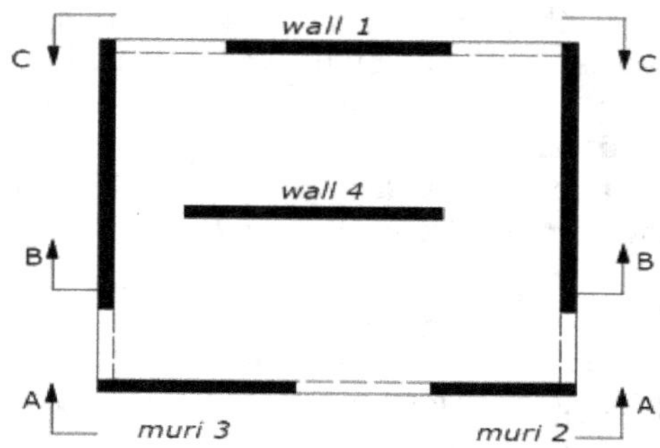

Figure 7.5.1 *- Fictitious example building (Planimetry)*

It is assumed that a wall carries only shear forces about its strong axes; the shear capacity for weak axes is neglected.

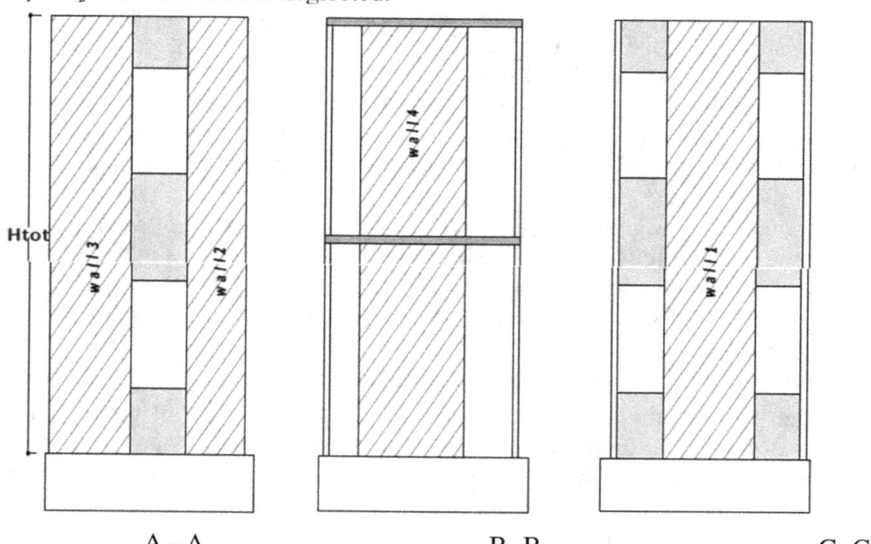

Fig. 7.5.2 *- Fictitious example building (Image)*

Assuming that the floors are completely rigid in their plan, providing equal displacement of the walls at floor levels, the capacity curve of the building in one direction can be obtained by superimposing the capacity curves of all the walls acting in this direction:

$$Vb(\Delta) = j\sum V j\ (\Delta)$$

(7.3)

j is the wall index, j = 1…n , where n is the total number of walls acting in one direction. This is permissible as long as the building geometry is relatively regular and the effects of torsion can be neglected

Figure 7.5.1 shows the plan and 7.5.2 three elevation views of a fictitious example building. Taking the x-direction into account, the four walls acting in this direction can be identified, namely wall 1, wall 2, 3 and 4. The contribution of the two walls in the y-direction is neglected. Wall 3 and wall 2 lie in a plane that constitutes a plane wall of the building (the wall's plane facade). Wall 1 constitutes the second plane wall of the building (also a plane facade wall).

Wall 4 constitutes a third plane wall inside the building. Also given are three elevation views of the buildings along the axes A-A, B-B and C-C. In the two outer plane walls, which constitute the two facades (A-A and C-C), the spandrels are somewhat deep, producing a considerable coupling effect, while in the inner plane (B-B) the wall is "connected" only by the floors, leading to a very reduced coupling effect.

The corresponding capacity curve as shown in Figure 7.7 is given by:

$$V_b(\Delta) = V_1\ (\Delta) + V_2\ (\Delta) + V_3\ (\Delta) + V_4$$

(Δ) (7.4)

Using a linear approximation of the capacity curve of the fictitious building taken as an example, the stiffness k of the linear elastic part corresponds to the sum of the effective stiffnesses of the walls;

$$k = \frac{V_{bm}}{\Delta_{by}} = \sum_j k_{effj}\,.$$

(7.5)

V_{bm} is the shear capacity and Δ_{by} is the maximum displacement the building can undergo without losing its integrity (Figure 3.8). In the case shown in Figure 7.5.1 this leads to a building stiffness k:

"

$$k = k_{eff1} + k_{eff2} + k_{eff3} + k_{eff4}$$

(7.6)

For this reason, to determine the capacity curve of the building, the capacity curve of the walls must first be determined.
For the capacity curve of a wall, a bilinear approximation is used, with a linear elastic part, up to the point where the shear capacity of the wall is reached, with a zero stiffness plastic part, completely (Fig. 7.6).

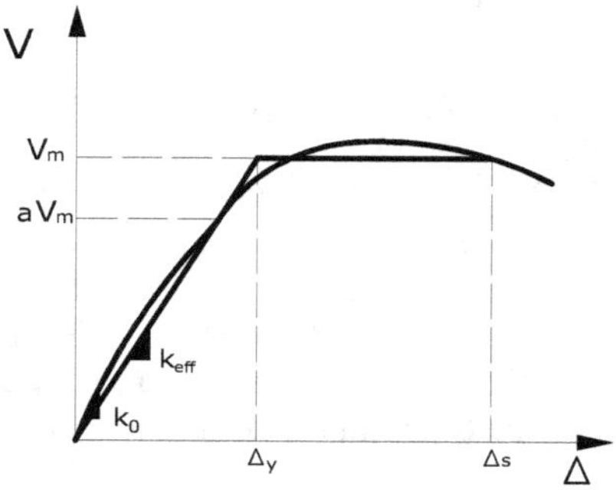

Fig 7.6- *Capacity curve of a single wall*

Thus, the bilinear capacity curve of a wall is determined by three parameters,

- the shear capacity V_m of the wall,
- the nominal yield displacement at the top of the wall Dy and
- the nominal ULTIMATE displacement at the top of the wall Ds

They are determined depending on the wall material, masonry or reinforced concrete.

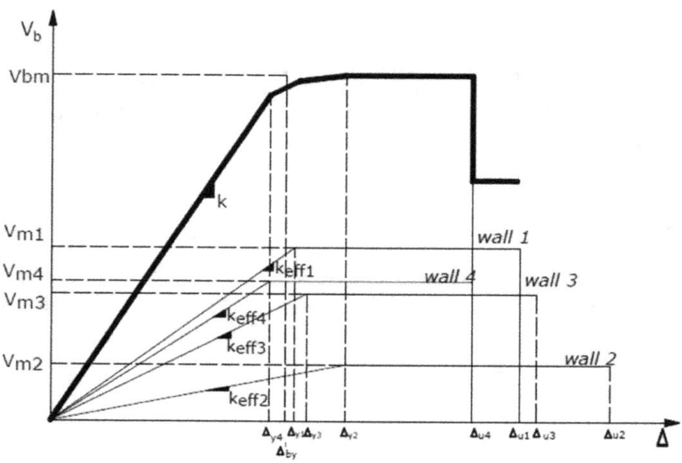

Fig 7.7- *Capacity curve for fictitious example building (fig. 7.5)*

7.3 - SEISMIC DEMAND [10]

The seismic demand is determined using a response spectrum. A response spectrum represents the **maximum response** of a single-degree-of-freedom (SDF) system as a function of its frequencies. Traditionally, in earthquake engineering, the acceleration response spectrum has been used for design based on forces and evaluation procedures. Recently, design and evaluation procedures have focused more on displacements and strains, which are considered to be the most important parameters. The use of a displacement response spectrum seems more appropriate. However, except for very low frequencies (f < 0.2Hz) the following simple relationship holds:

$$S_a \approx \omega^2 * S_d$$

(7.7)

where Sa and Sd are the acceleration and spectral displacement respectively and $\omega = 2\pi f$ is the corresponding circular frequency, (f is the frequency in Hz).

It should be noted that although the response spectrum is considered to be valid for an entire region, it depends on local field conditions and can vary considerably even within a region. This effect is taken into account in the earthquake design scenario through microzonation.

Recently the use of the spectral response in the ADRS format has become increasingly popular. As the name suggests, this is a diagram where the *spectral acceleration is plotted against the spectral displacement.* The lines emanating from the origin of this diagram have constant periods.

The advantage of this format is that capacity and demand can be plotted on the same diagram, allowing for a graphical solution. However, it should be noted that this is only a different representation of the same data, it does not provide any further information. The use of both formats is, however, at the discretion of the engineer. In the following, the displacement response spectrum will be used to present seismic data throughout this work.

The use of the response spectrum assumes that the building, which can be viewed as a multi-degree-of-freedom system (MSFS), where the masses are concentrated at the floor levels and the wall mass is distributed between the two levels, above and below (Figure 3.11), can be described by an equivalent single-degree-of-freedom system (SFS) characterized by an equivalent mass and an equivalent stiffness, having the same fundamental frequency as the MSFS:

$$f_1 = \frac{1}{2\pi} \cdot \sqrt{\frac{k_E}{m_E}}$$

(7.8)

If the stiffness **k** of the real structure, obtained from the linear approximation of the building capacity curve (Figure 7.5.1 and Equation 7.5) is used as the equivalent stiffness k_E of the (SFS) system:

$$k_E = k = \frac{V_{bm}}{\Delta_{by}} \qquad (7.9)$$

The equivalent mass is given as:

$$m_E = \Sigma\, m_i\, \Phi_i \qquad (7.10)$$

where $\mathbf{m_i}$ is the concentrated mass and Φ_i is the first mode displacement of the i-th floor, normalized to the first mode displacement at the highest floor, $\Phi_n = 1$. The equivalent height is:

$$h_E = \frac{\sum h_i m_i \phi_i}{\sum m_i \phi_i} \qquad (7.11)$$

in which $\mathbf{h_i}$ is the height of the i-th floor.

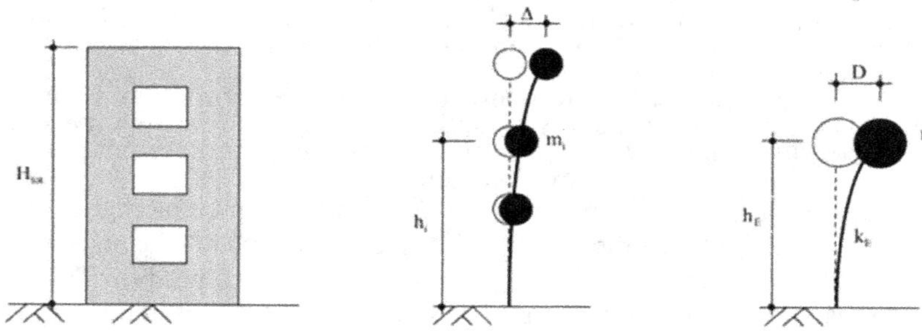

MDF

SDF

Fig. 7.8 The equivalent SDF system of the one with MDF

Any value of the MDF system can be transformed to an equivalent SDF system, using the following equation:

$$Q = \Gamma \cdot Q_E \,. \qquad (7.12)$$

Q - represents the values in the MDF system (shear force at base $\mathbf{V_B}$, displacement at top **D**) Q_E- represents the values in the equivalent SDF system (force F_E, displacement **D**, with maximum displacement denoted as S_d).
Γ- is the modal participation factor defined as:

$$\Gamma = \frac{\sum m_i \phi_i}{\sum m_i \phi_i^2} \,. \qquad (7.13)$$

Two different approaches exist to obtain the displacement demand at the top of the building, taking into account the nonlinear behavior of the building. One is to use the inelastic demand spectrum, the other is to use the elastic large-damping spectrum. Using the inelastic demand spectrum, the displacement demand at the top of the building $\mathbf{D_D}$ (n-th floor), in terms of the equivalent elastic displacement, is:

$$\mathbf{D_D = c_n * D_{be}} \tag{7.14}$$

This is illustrated in **Fig**. 7.9, where the base shear force–top displacement relationships are shown for linear elastic and nonlinear behavior. The constant $\mathbf{c_n}$ can be defined as a function of the resistance reduction factor R and the ductility requirement μ_D:

$$c_n = \mu_D / \mathbf{R} \tag{7.15}$$

μ_D is defined as:

$$\mu_D = \Delta_D / \Delta_{by} \tag{7.16}$$

And the resistance reduction factor R can be represented as:

$$R = \mathbf{V_{be} / V_{bm}} \tag{7.17}$$

With: $V_{be} = k * \Delta_{be} = k * \Gamma * S_d(f_1)$
(7.18)

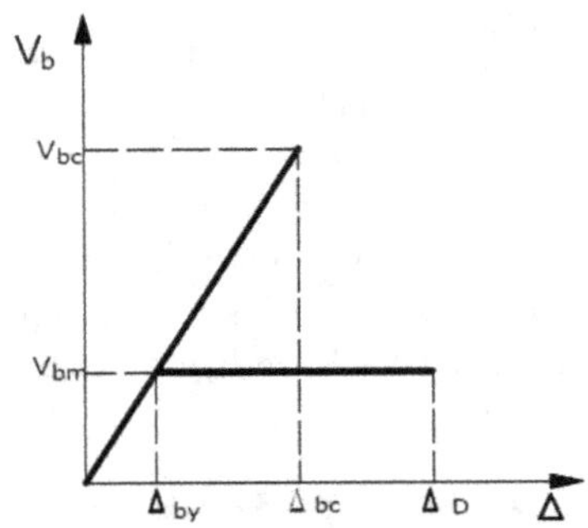

***Fig. 7.9-** The relation between shear force at the base -displacement at the top, for linear elastic behavior and nonlinear behavior.*

The first to study this type of relationship was Newmark et al. [VN 60] [VNC 65]. Based on the elastic and inelastic response spectra of the May 18, 1940 earthquake, they observed that:

i) in the low-frequency range, an elastic system and an inelastic system have approximately the same maximum displacement;

ii) in the intermediate frequency range, the principle of conservation of energy gives a fairly good approximation, i.e. the area under the base shear force-top displacement diagram of the linear elastic system is equal to the area under the base shear force-top displacement diagram of the nonlinear system.

iii) in the high-frequency range, the elastic system and the inelastic system have the same force.

In a simplified way, these results can be presented as a function of frequency; The principle of maximum equal displacement, valid for $f_1 < fc_2$:

$$R = \mu D \tag{7.19}$$

ii) The principle of equal energy, valid for f1 > fc2:

$$R = \sqrt{2\mu_D - 1} \tag{7.20}$$

iii) f1 > 33Hz: R=1 (7.21)

The third case is usually not applicable to buildings. Both the limiting frequencies f1 and f2 depend on the ductility of the system and the characteristic value of the elastic spectrum. Typical values are:

$$f1 = 1.4 \text{ Hz dhe } f2 = 2.0 \text{ Hz.} \tag{7.22}$$

For frequencies between f1 and f2, R is obtained by interpolation between these two principles. These results should not be considered "exact", but rather as indicative of an average trend. However, discrepancies are not uncommon. By finding the constant cn the required displacement **D_D** at the top of the building (n-stories), can be obtained from the following equation:

$$\mathbf{\Delta_D = c_n * \Gamma * \varphi_n} * S_d(f_1) \tag{7.23}$$

ϕ_n corresponds to the displacement of the first mode at the top floor (n-th floor) of the MDF system. Other **R-μ_D- f_1** relations have been developed, based on the same principle.

Vidic et al. [VF 94] proposed the following R-μ_D- f_1 relation:

$$R = \begin{cases} 1.35(\mu_D - 1)^{0.95} * \frac{T_1}{T_0} + 1, & T_1 \leq T_0 \\ 1.35(\mu_D - 1)^{0.95} + 1, & T_1 > T_0 \end{cases}$$

(7.24)

$$T_0 = 0.75\mu_D^{0.2} \cdot T_c \leq T_c$$

T_c is the characteristic period of the ground motion and is usually defined as the transition period from the constant acceleration segment of the response spectrum to the constant velocity segment of the spectrum.
T_1 is the fundamental period of the building, T1 = 2π/f1
Many other R– μD–f_1 relationships exist, a good overview is given by Miranda and Bertero [MB 94] and Chopra and Goel [CG 99]. It is often considered a major drawback of all methods that they do not account for the change in fundamental frequency with increasing nonlinear behavior, or hysteretic characteristics of energy dissipation. As damage increases, stiffness decreases and this will directly affect the fundamental frequency (and thus the spectral response) and increase the damping. Also, focusing on displacement as an important parameter, it is felt that:

- The inelastic spectrum approach focuses heavily on strength.
- In the second approach, based on the replacement structure approximation of Shibata and Sozen [SS 76], the displacement requirement at the top of the building is found from a high-damping elastic spectrum with equivalent stiffness corresponding to the shear stiffness:

$$k_{equ} = \frac{V_m}{\Delta_D}. \qquad (7.25)$$

The equivalent stiffness depends on the required displacement at the top and illustrates the fact that the capacity of a building and the seismic demands are not independent. It should be noted that the equivalent stiffness corresponds only to the last point of the response, regardless of the initial stiffness and the change in stiffness along the loading path.
The critical point of the procedure is the use of an elastic spectrum with very high damping.
Eurocode 8 [EC8] proposes the following correction factor for the response spectrum for damping values different from 5%, which is often adopted in studies on displacement-based design [Ca 99] [BCA 01]:

$$\eta = \sqrt{\frac{7}{2 + \beta_{equ}}}. \qquad (7.26)$$

β_{equ} - is the equivalent viscous damping which corresponds to a combination of the viscous damping inherent in the structure (usually taken as 5%) and the hysteretic damping, which is related to the area inside the hysteresis loop and therefore depends on the ductility requirement. The most common way to define the equivalent viscous damping is to equate the energy dissipated (lost) in a hysteresis loop of the structure to the energy of an equivalent viscous system [Ch 95] resulting in:

$$\beta_{equ} = \frac{1}{4\pi} \cdot \frac{E_D}{E_{s0}} + \beta_v . \tag{7.27}$$

-
- E_D is the energy dissipated (absorbed) by the damping corresponding to the area inside the hysteresis loop
- E_{s0} is the maximum tensile energy
- β_v is the viscous damping corresponding to the elastic response.
- Shibata and Sozen [SS 76] proposed the following relation to calculate the equivalent viscous damping based on dynamic tests of reinforced concrete elements and single-story frames:

$$\beta_{equ} = 20 \cdot \left(1 - \frac{1}{\sqrt{\mu_D}}\right) + \beta_v . \tag{7.28}$$

Figure 7.9 shows a comparison of different approaches, where the required displacement at the top of the building is given as a function of the peak ground acceleration. The effects of nonlinearity are taken into account, using the principle of equal energy, the R–μD – f1 relation proposed by Vidic et al. [VF 94] and the surrogate structure approximation.
A discussion of these approaches is beyond the scope of this paper. The engineer should be aware that different approaches exist, however, for the purposes of earthquake scenario design, the difference in results is of little importance and consequently the first approach, which uses the equal energy principle and the equal displacement principle, provides a straightforward formulation of the required displacement at the top, which will be used in the following.

7.4- The vulnerability FUNCTION

By varying the “intensity” of the seismic demand, through the continuous increase of the spectral displacement, from zero onwards, the demand for building displacement increases continuously, following equation (7.28) the curve **Sd(f1)** - Δ is obtained (Figure 7.10). However, this is not yet a function of vulnerability. The displacement at the top should be associated with a damage limit. Different approaches exist, often using quantitative measures, where the damage is expressed as a percentage of the total destruction [ATC 13] or as a fraction of the ultimate deformation capacity [FG 96]. It is felt by the author that these quantitative measures are not very suitable for earthquake scenario projects, where the interest is more related to monetary losses and casualties. A qualitative description of the damage is therefore suggested, based on the damage classification proposed by the European Macroseismic Scale [EMS 98], which divides the damage into 5 levels, ranging from negligible damage to destruction. A copy of this damage classification is given in Appendix A1. In order to use these damage scales, the "indicators" must be defined as defining a point on the building capacity curve at which the object enters the next damage class. This is shown in Figure 7.11 with reference to the capacity curve of the imaginary building – example of Figure 7.1.

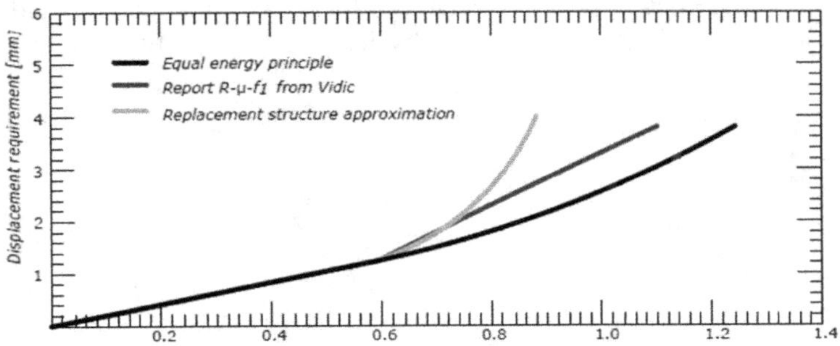

Fig. 7.10- *Comparison of different approaches to consider the effects of nonlinearity*

The main parameter used as an indicator is **structural damage**, by checking individual walls as well the entire building.
By showing the damage levels from Figure 7.11 in the Sd(f1)–Δ relationship, the vulnerability function is taken from Figure 7.12. The vulnerability function is linear for Δ< Δby, since the building's capacity curve for Δ < Δby is in the elastic linear zone (Figure 3.14), and from this, it follows that cn = 1 in equation (7.23). For Δ> Δby, the building's capacity curve enters the plastic region, and therefore cn = μD / R in equation (7.23). For buildings with f1 ≥ fc1, the vulnerability function becomes nonlinear for Δ> Δby.

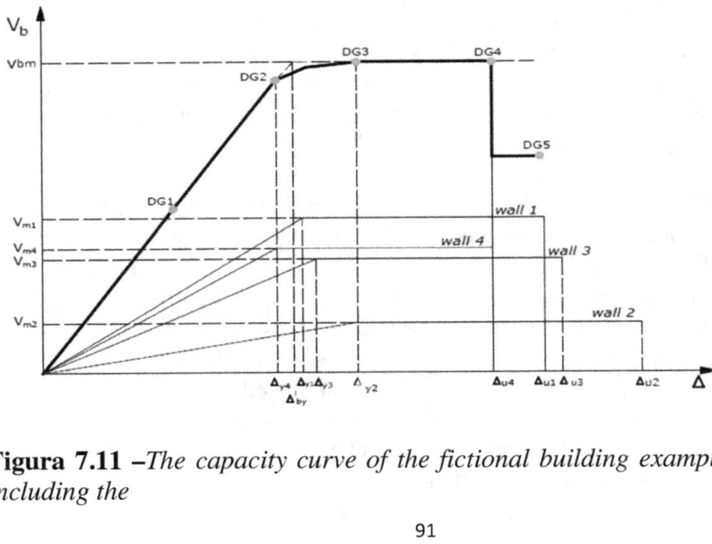

Figura 7.11 –*The capacity curve of the fictional building example in Fig. 7.8, including the*

damage levels.

The use of these damage scales allows a "visual" interpretation of it, depending on the object and local practices. Victims are also a consequence of the physical condition of the building.

The use of these damage scales allows a "visual" interpretation of it. Depending on the object and local practices, financial losses (both structural and non-structural) and victims are a result of the physical condition of the building.

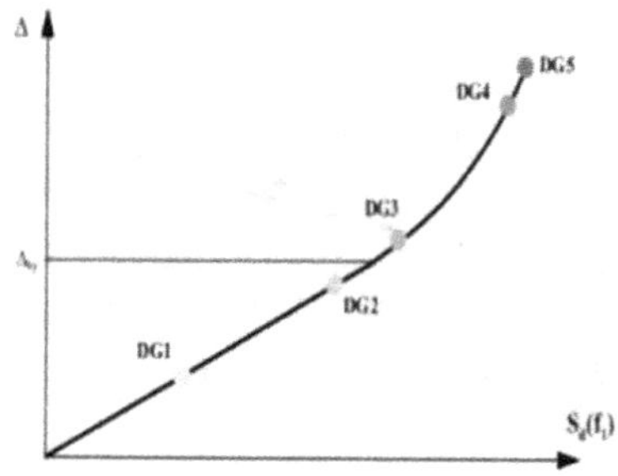

Figura 7.12: Vulnerability function for the fictitious building example of Fig. 7.1

7.5 VULNERABILITY OF MASONRY BUILDINGS

7.5.1 Capacity curve of brick masonry

The bilinear capacity curve of a wall is determined by three parameters:

- the maximum shear force of the wall Vm,
- the nominal yield displacement at the top of the wall Δy and
- the nominal ultimate displacement at the top of the wall Δu.

The next three sections demonstrate how these three parameters can be defined for a brick wall. Although masonry structures do not yield in the following, the transition point between the linear elastic and perfectly plastic zones of the bilinear capacity curve will be referred as the "yield point", in analogy with reinforced concrete structures.

By definition (Fig. 7.3), a wall consists of several piers per floor, separated by relatively rigid zones. The capacity of a wall is determined by the capacity of the pier that fails first. The joint regions, as well as the spandrels, are not taken into account, assuming that they can accept the internal forces that are necessary for the equilibrium conditions. As already mentioned, this is not really the case for the spandrels, which will usually not be able to support the forces required for equilibrium at the yield points of the first pier, undamaged, but which has suffered cracks. This can be considered in a very simplified way, assuming a reduced stiffness of the spandrel and taking into account a reduced coupling effect, due to the formation of early cracks in the spandrel.

A more complex model can take into account the spandrel capacity. Therefore, to calculate the capacity curve of a wall, the capacity of a pier with height hp and length lw is first considered. Figure 7.12 is essentially a repetition of Figure 7.4, where the wall element is now a pier.

Due to the coupling effect, the spandrel at a height h0 will have a moment of zero (see the moment distribution of the frame in Figure 7.4). The dependence of h0 on the stiffness of the spandrel is discussed in Section 3.5.3.

A more complex model may take into account the capacity of the spandrel. Therefore, to calculate the capacity curve of a wall, the capacity of a pier with a height $\mathbf{h_p}$ and length $\mathbf{l_w}$ is initially considered. Figure 7.12 is essentially a repetition of Figure 7.4, where the wall element is now a pier. Due to the coupling effect, the spandrel at a height h_0 will have zero moment (see the moment distribution of the frame in Figure 7.4). The dependence of h_0 on the stiffness of the spandrels is discussed in Section 3.5.3.

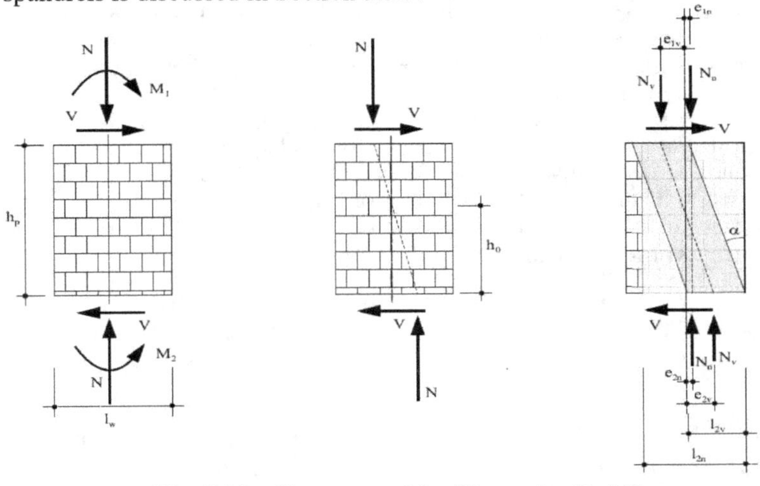

***Fig. 7.12**- Geometry of the Pier + Applied Forces*

It is assumed that h0/h behaves similarly h0/hst, and thus the values from Figure 3.6 can be used. h0 is always measured from the bottom of the respective element. In the case of the end zones of the connection, this leads to an error that increases with the growth of the dimensions of the joint zones and with the reduction of the coupling effect. Therefore, a deep spandrel, and thus an increase in the joint dimension, causes an increase in the coupling effect, and vice versa. The error will be small and, for this reason, has been neglected in the following analysis.

In the case shown in Figure 4.7, h0<hp indicates an extremely strong coupling effect. To account for the reduction in the coupling effect due to the formation of early cracks in the spandrel, a 50% reduction in the stiffness of the spandrel has been proposed. For a given value of h0, M1 and M2 can be expressed in terms of V:

$$M_1 = V^* (h_0 - h_p) \qquad (7.29)$$

$$M_2 = V^* h_0 \qquad (7.30)$$

For $h_0 < h_p$, the bending moment in the pier changes its sign, as in the case of Figure 7.12. For $h_0 > h_p$, the bending moment does not change its sign. This corresponds to the case in Figure 7.4.
By taking M_1 and M_2 from equations (7.29) and (7.30), the lower-bound theorem of plasticity can be used to calculate the shear capacity of the pier for h=hp. The masonry pier that reaches its limit state first, under the given force distribution, determines the shear capacity of the entire wall. In most cases, this is the ground-floor pier. However, in cases where there is a significant variation in stiffness and mass along the height of the building, a pier located at mid-height may reach the limit state first.
As explained in Section 7.5.4, a bilinear approximation has been used for the capacity curve of a wall (Figure 7.6), with a linear elastic segment up to the point where the wall's shear capacity Vm is reached, and a perfectly plastic segment with zero stiffness thereafter.
In reality, the response of masonry is predominantly nonlinear; as damage due to cracking increases, masonry simultaneously undergoes degradation in both strength and stiffness. However, for practical purposes, a bilinear approximation is often used to idealize the shear behavior of brick masonry.
The effective stiffness of the linear elastic segment can be determined using a secant stiffness at **αVm**, with $0.6<\alpha<0.750.6$, and selecting Vm, such that the bilinear curve is equivalent to the experimental curve in an energy sense [CM 97]. Krahasimet me rezultatet eksperimentale [GT 84], tregojnë se në këtë rast një faktor reduktimi i ngurtësise prej 0.5 në 0.7 është i përshtatshem:

$$\mathbf{k_{eff}} = (0.5 \div 0.7) k_0 \qquad (7.31)$$

Figure 7.13 shows the elastic displacement shape in a six-story masonry building due to a triangularly distributed horizontal force, using frame analysis. It can be observed that the displacement at the lower part of the building **is less** than what would be expected for a frame structure. This is because the walls in a masonry structure **are very massive**, making the construction conditions at the base of the building more significant compared to frame structures made of concrete, where the columns are relatively slender. Additionally, the displacement is given using a constant drift along the height of the building, equal to the drift at the first story, and it can be observed that this simple approximation provides good results for the displacement at the top of the building.

k_0 - is the stiffness of the uncracked wall.

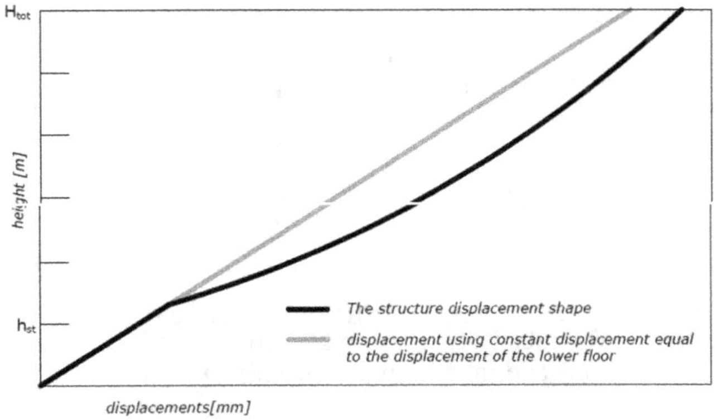

***Fig.7.13**- The displacement diagram of a six-story masonry building due to a triangularly distributed horizontal force.*

Using the assumption of continuous drift **δ** over the height of the building **Htot** , the elastic displacement Δ at the top of a wall can be expressed as follows:

$$\Delta = \delta * H_{tot} \qquad (7.32)$$

The drift **δ** can be calculated using the principle of virtual work. Considering a pier with a height **hp**, with the zero-moment point at a height h_0, the bending moment and distributed shear force due to **(a)** the actual forces and **(b)** a virtual unit force are shown in Figure 7.14. It is assumed that the first-story drift is equal to the drift in the pier. In reality, the displacement above the first floor is not uniform. Piers, being the weakest part of a wall, will deform more, whereas the joint areas are relatively stiff and will deform less. Therefore, the assumption of a continuous drift equal to the pier's drift overestimates the linear elastic deviation. However, for the purposes of this assessment method, the results are quite accurate (see the comparison with experimental results in Section 4.4).

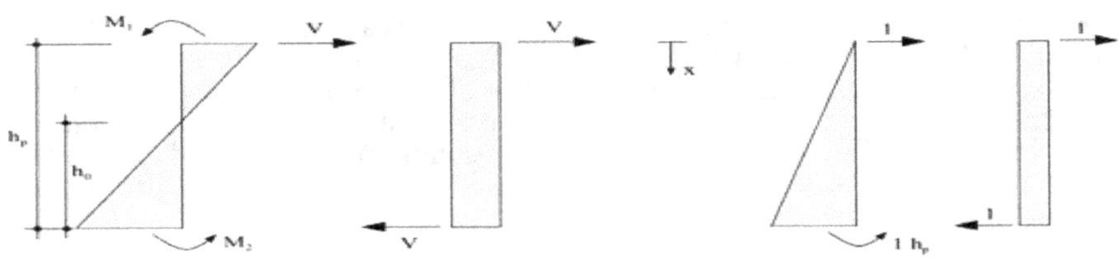

a) Mx dhe Vx b) mx dhe vx

Fig. 7.14 The distribution of the bending moment and shear force at the base of the wall pier due to: a) the actual forces and b) a virtual unit force

The horizontal deformation at the top of the pier in the direction of the unit force can be calculated as:

$$d = h \int_0^{hp} \frac{M(x)*m(x)}{EI_{eff}} hdx + \int_0^{hp} \frac{K*V(x)*v(x)}{GA_{eff}} dx \qquad (7.33)$$

$$M(x) = M_1 + V \cdot x; \quad V(x) = V; \quad m(x) = x \text{ and } v(x)=1$$

with:

EI_{eff} dhe GA_{eff} are the effective bending stiffness and effective shear stiffness of the pier, respectively. By substituting the expressions for M(x), V(x), m(x), and v(x) into Equation (7.33) and integrating from zero to hp, the displacement at the top of the pier is obtained.

$$d = \frac{V*h_p^3}{3*EI_{eff}} + \frac{M_1*h_p^2}{2*EI_{eff}} + \kappa * \frac{V*h_p}{GA_{eff}} \qquad (7.34)$$

k- It is a shape factor that depends on the specific form of the cross-section. For a rectangular cross-section, k=6/5.

By substituting Equation (7.29) for M_1 and dividing by the height hp of the pier, the following expression for drift hp is obtained:

$$\delta = V * \left(\frac{h_p*(3h_0-h_p)}{6*EI_{eff}} + \frac{V*h_p}{GA_{eff}} \right) \qquad (7.35)$$

The displacement **Δy**, at the yield displacement threshold, at the top of the wall, can be determined as the displacement for V=Vm, using the equations (7.35) dhe (7.32):

$$\Delta_y = V_m * H_{tot} * \left(\frac{h_p*(3h_0-h_p)}{6*EI_{eff}} + \frac{\kappa}{GA_{eff}} \right) \qquad (7.36)$$

Although Figure 7.14 shows the bending moment and shear force distribution for $h_0<h_p$, Equation (7.35) is also valid for the case where $h_0>hp$, since the boundary conditions are fully determined by h_0.

- The point at which the shear strength in the wall is reached does not necessarily mean the failure of the wall. Unreinforced masonry should not be considered brittle; in fact, measurements of the behavior of wall elements have shown that unreinforced masonry can possess a considerable capacity for plastic deformations. Plastic deformations as much as eight times the yield strain have been observed! Based on a perfectly elastic-plastic behavior, the ultimate displacement can therefore be expressed as:

$$\Delta_u = \mu_w * \Delta_y \qquad (7.37)$$

μ_w - is the ductility of the wall. In the next section, the ductility of masonry will be discussed

7.5.2 DUCTILITY of masonry structures

The definition of ductility is based on a linear elastic-perfectly plastic behavior and is defined as the ratio of the total elasto-plastic deformation to the elastic deformation at yield:

$$\mu = u / y \tag{7.38}$$

In the case of reinforced concrete structures it is common to distinguish between local ductility [curvature ductility, torsional ductility] and system ductility [displacement ductility]. For unreinforced masonry it is convenient to distinguish between the following ductilities:

i) Displacement ductility of a wall element μ_{WE}
ii) Displacement ductility of a wall μ_{W}
iii) Displacement ductility of the building μ_{Δ}

Ductility is reduced from i) to iii): $\mu_{WE} > \mu_{W} > \mu_{\Delta}$

- The ductility of unreinforced masonry is not ductility in the traditional sense such as the ductility of reinforced concrete, which results from the plastic deformation of the reinforcing steel. Here, the ductility according to definition (4.27) is obtained from the relative slip of the wall elements during cracking without significant loss of strength. It is a function of the normal acting stress σ_n, of the geometry expressed by the ratio h/l_w, of the material properties, expressed by the compressive strength f_{mx} and the boundary conditions bc:

$$\mu = \mu\{s_n, h/l_w, f_{mx}, bc\} \tag{7.39}$$

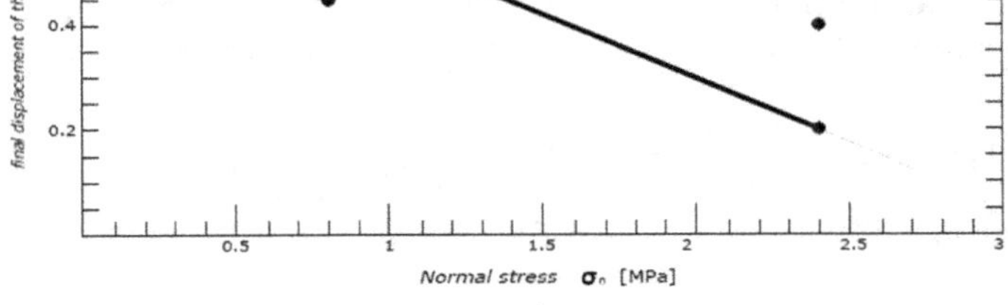

Fig. 7.15 **-** The relationship between the ultimate drift and the normal stress of the wall elements, tested in *ETH Zürich*

The dependence of ductility on normal stress has been quantitatively determined using the results of masonry elements tested at ETH Zürich by Ganz [GT 84].

Using a linear interpolation of the test results (Figure 4.10), the ultimate drift of a wall element, **δu** (in [%]) is determined as a function of the normal stress **σn** (in [MPa]) acting on the wall element:

$$\delta_u = 0.8 - 0.25\ \sigma_n \qquad (7.40)$$

The ductility of the wall element is then obtained from:

$$\mu_{WE} = \delta_u / \delta_y \qquad (7.41)$$

Determining the ultimate drift first is more reliable, because the value of **μWE** depends on the yield point, which in turn involves high uncertainties.
Equation (7.40) illustrates the fact that for small axial forces cracks usually pass through horizontal joints (bed joints), according to a diagonal pattern; separate parts of the wall can slide over each other resulting in large relative deformations, without significant loss of strength.
For high axial forces, cracks pass through the brick units and as a result the separated parts of the wall tend to slide with small deformations downwards, along the more regular surfaces of the cracks. However, the tests were all carried out with test specimens of the same geometry (walls of the same height, length and thickness) and with the same compressive force, Fmax=8.25MPa, normal to normal to the bonding mortar, neither the influence of geometry nor the influence of compressive strength on ductility could be studied.

The influence of geometry was revealed by static cyclic tests on masonry test specimens with different aspect ratios carried out at the Joint Research Centre of the European Commission, Ispra, Italy [AMM 94].
The results show that in wall elements with high aspect ratios, the shaking mode leads to high drift capacities, without significant degradation of resistance.
Wall elements with low aspect ratios, in the meantime, tend to fail in shear, leading to reduced drift capacities. To take into account the different behavior of masonry elements with different aspect ratios, it has been suggested to increase or decrease the ultimate drift capacity of a pier, in the following simplified way:

$$\delta_u = \begin{cases} 0.8 * (0.8 - 0.25\sigma_n) & \frac{h_p}{l_w} < 0.5 \\ 0.8 - 0.25\sigma_n & 0.5 < \frac{h_p}{l_w} < 1.5 \\ 1.2 * (0.8 - 0.25\sigma_n) & \frac{h_p}{l_w} > 1.5 \end{cases} \qquad (7.42)$$

This, however, needs further investigation, especially considering the energy dissipation in the shaking mode and its implications for the resistance reduction factor R.. An upper limit to the ductility of a wall element is set at $\mu_{WE}max = 12$.

In many cases, where the walls are quite thin, it is usually the pier end that first reaches the limit conditions and enters the plastic state. Assuming that the upper part of the walls as well as the spandrels remain elastic, a sway pier mechanism is formed (section 5.4.3). This is illustrated in Figure 7.16 where we have

$$\Delta_u = \Delta_y + (d_u - d_y) \qquad (7.43)$$

d_y - is the displacement at the top of the wall, at the yield point, Δy is the displacement of the wall at the top, at the yield point of the lower part of the pier (which corresponds to the yield point of the entire wall), du is the ultimate deformation of the lower part of the pier:

$$d_u = \mu_{WE} \cdot d_y \qquad (7.44)$$

Thus, the ductility for wall displacement can be assumed to have a linear elastic displacement form:

$$\mu_{WE} = 1 + hp/Htot * (\mu_{WE} - 1) \qquad (7.45)$$

The acceptance of a pier impact mechanism may lead to a somewhat conservative estimate of the ultimate displacements. A partial plasticity mechanism of the spandrel seems to be considered valid. However, this requires consideration of the spandrel capacity which is not considered in this study.

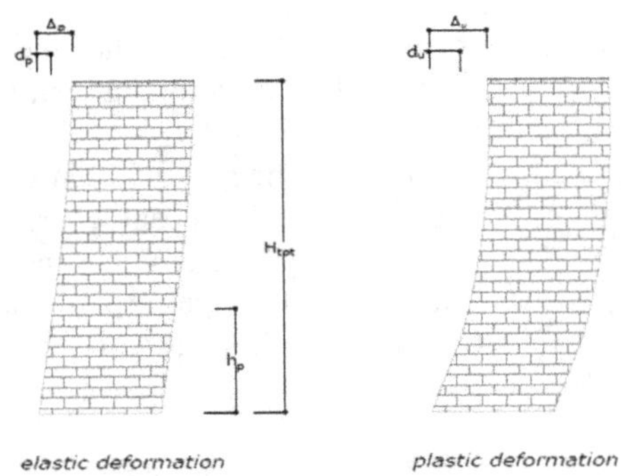

Fig 7.16- Deformation of brick masonry

In the case of short walls subjected to small normal forces, the upper part of the wall reaches the boundary conditions first, which are determined by sliding along the mortar layer

7.5.3 MASONRY QUALITY

When a detailed assessment of an existing individual building is necessary, for example, to determine reconstruction strategies, the exact properties of the masonry must be understood. Various non-destructive and slightly destructive tests exist that can be applied on-site to quantitatively assess the critical performance parameters of masonry.

Non-destructive tests include thermography and ground-penetrating radar, which are used to identify damaged joints and voids, as well as sonic tomography, which uses wave velocity as an evaluation of material resistance.

Partially destructive methods include:

- Single tests using a flat jack for measuring stresses and modulus of elasticity on-site.
- Double tests using a flat jack for measuring the strength of materials.
- Penetration tests to measure the shear strength of horizontal joints (bed joints) on-site.

On-site tests for assessing critical parameters entail high costs, both financial and temporal. Their validity is limited to the specific building being considered. Therefore, they are not very suitable for earthquake scenario projects.

Following this, a very simple method is described for determining the properties of masonry based on the year of construction. This is a rough assessment based on older documents, aiming to capture the general trend in masonry execution. For an individual building, this assessment may lead to very inaccurate results; however, on average, the properties of masonry should be assessed with a reasonable degree of accuracy for earthquake scenario projects. In such cases, a large number of buildings need to be evaluated, making the average behavior of a group of buildings more relevant than that of an individual building.

Unlike other materials, masonry is characterized by a wide variety of its constituent materials and the significance of their interaction on its properties. Although significant progress has been made in recent years in the field of composite modeling, there is currently no reliable analytical model that directly links the properties of the constituent materials (stone and mortar) with the resulting properties of masonry. However, experimental tests on masonry units in the laboratory have provided valuable information on the influence of various parameters and their priorities.

In the following, we have focused on regular masonry using clay bricks (briefly referred to as brick masonry), as this is the primary material for residential buildings. Although one of the oldest and most widely used construction materials in the world, the use of brick masonry has been poorly documented, and projects have been based on experience and very rough guidelines. Research is somewhat limited; interest has only increased in the past few decades, and a standardized procedure for determining the properties of brick masonry has not existed for quite some time.

The material requirements regarding the compressive strength of brick units, the mortar used and its main strength, as well as the allowable stresses, are presented in Table 7.1

Tab. 7.1*- Requirements for brick masonry materials according to SIA 113,* ***1943*** **[10]**

	Compressive strength of brick units	Permitted stresses		
		Cement mortar	Llac gelq.-cement	Lime-cement
Normal brick unit	22	1.4	0.9	0.6
High-grade brick	35	2.2	1.3	0.8

Tab. 7.2*- Requirements for brick masonry materials according to SIA 113,* ***1965*** **[10]**

	Compressive strength of brick	Nominal resistance of	
		Cement mortar f_m	Lime-cement
Normal brick unit	15	-	3.0-5.5
High-grade brick	25-35	8.5-15	5.0-9.5
Special brick unit	40	16-25	-

Tab. 7.3*- Requirements for brick masonry materials according to SIA 177,* ***1980*** **[10]**

	Compressive strength of brick	Nominal resistance of	
		Cement mortar f_m	Lime-cement
Normal brick unit	20	6.0-10	3.0-5.5
High-grade brick	25-35	10-15	-
Special brick unit	40	16-23	-

Tab. *7.4- Requirements for brick masonry materials according to SIA 177,* ***1995*** **[10]**

	Compressive strength f_b [MPa]	Nominal resistance of	
		Compressive strength of	Compressive strength
Brick masonry (BM)	28	15	8
Light brick masonry	14	5	4

CHAPTER 8- PUSHOVER ANALYSIS FOR SEISMIC ASSESSMENT OF MASONRY BUILDINGS. [8]

8.1 - NONLINEAR STATIC ANALYSIS OF MASONRY BUILDINGS

The model described here, Simplified Analysis of Masonry, (abbreviated SAM), was conceived for the global analysis of new and existing masonry buildings, in which the resisting mechanism is governed by the in-plane reaction of the walls. The collapse mechanism due to out-of-plane dynamic reaction is not considered in the model, and will be evaluated in a separate modeling.

Global seismic analysis of a masonry building is meaningful if appropriate means, such as ties or beam webs, prevent local and global out-of-plane collapse, which would otherwise occur prematurely even at low seismic intensity. This model was first developed for planar structures [Magenes and Della Fontana,1998], and later extended to three-dimensional buildings [Magenes, 1999]

Considering a multi-story brick wall, loaded in plan by horizontal forces, if the geometry of the openings is sufficiently regular, it is possible to idealize it as an equivalent frame, composed of vertical elements (piers), beam elements (spandrels), and connection joint elements (Fig. 1). The vertical elements (piers) and horizontal elements (spandrels) are modeled as beam-column elements, with shear deformation, while the joint elements are assumed to be infinitely resistant and rigid, and are modeled as rigid connections at the ends of the spandrel and pier elements.

The pier element is assumed to have an elasto-plastic behavior, with limited deformation. The element exhibits a linear elastic behavior until one of the possible failure criteria is met. The elasto-plastic idealization approximates the envelope of experimental resistance under cyclic loading. The following failure mechanisms are predicted:

Failure by bending or "swinging" occurs when the moment **M** at any end of the effective length of the pier reaches the ultimate moment ***Mu***, which is a function of the axial force, the geometry of the section, and the compressive strength of the masonry ***fu***. A plastic hinge is created in the section when ***Mu*** is reached.

Diagonal shear cracking is defined as the lesser of the resistances associated with the failure of the mortar and the unit brick, as proposed in [Magenes and Calvi, 1997].

When the failure criterion is met, shear plastic deformation occurs as in Figure 2, where a limit **θu**, at the maximum chord rotation, is set, beyond which the resistance it is zoroed.

The chord rotation is expressed as the sum of the bending and shear deformations **$\theta=\varphi+\gamma$**, see (Fig. 3) and is a generalization of the concept of drift, for non-symmetrical boundary conditions of a pier, subjected to bending and shear. A suggested limit for unreinforced masonry is $\theta u= 0.5\%$
Shearing can occur at any end of the pier section, and is a function of the resistance of the (horizontal) bed joint and the flexural cracking in the section. An elastic deformation due to shearing is modeled in a similar way to the case of diagonal shear cracking. Full expressions for the resistance criteria can be found in [Magenes and Calvi, 1997] and [Magenes and Della Fontana, 1998]. The failure criteria are such that the flexural resistance is **non-zero** only in the presence of axial compression. Central tension **is not allowed**, i.e. the axial stiffness of the pier is zero for the central tensile deformation.

The spandrel element is formulated similarly to the pier element, taking into account the different orientation of the bed (horizontal) joints with respect to the axial force. The possible failure mechanisms are bending and shear. For the bending failure the formulation is identical to the pier element. For the shear resistance, it is considered that, due to the openings above and below the spandrel element, the horizontal joints have almost zero normal stress, and therefore the shear resistance is produced only by cohesion. The nonlinear behavior of the spandrel, collapsing in shear, is seen in Fig. 4, in which the degradation of the resistance is predicted for increasing values of shear strain.

From the meaning of the parameters **α,** γ_1 **,** γ_2 it is possible to obtain a variety of behaviors, from elastic brittle to elastic perfectly plastic. This basic hypothesis, most articulate, allows to take into account the tendency for a more brittle, post-peak behavior of spandrels, compared to piers, which has some importance in the results.

To analyze three-dimensional buildings, the plan model was extended [Magenes, 1999] by formulating the fundamental laws of piers and spandrels in three dimensions. An independent behavior of the pier or spandrel in two orthogonal, principal planes, parallel to the element axis, is assumed. The out-of-plane behavior is modeled in a similar manner to the in-plane behavior. Composite walls (e.g. walls intersecting orthogonally) are decomposed into simple walls with rectangular cross-section. If the intersecting walls are effectively connected, it is possible to simulate a connection that characterizes a suitable rigid offset and to impose a continuity of displacements at the ends of the rigid offset, at the floor levels.
An important issue is considered the possibility to model the presence of a concrete ring beams, whose role can greatly affect the connection between the piers. The ring beams are modeled as ELASTO-plastic frame elements, which can degenerate in bending, in plastic hinges. The steel connections can be modeled as ELASTO-plastic beam elements. The rigid diaphragm of the basement can be stimulated by imposing a kinetic constraint on the joints at the floor level.

8.2- METHOD VERIFICATION

The first applications of the method [Magenes and Della Fontana, 1998] were made on two- and three-story walls, comparing the results with those obtained from nonlinear finite element analysis of plane stresses, improved with a specific fundamental law for unreinforced brick masonry [Gambarotta and Lagomarsino, 1997].
In such analyses (an example is given in Figure 5) a very good agreement of the results of the two methods was found in terms of the general stress-rupture mechanism, proving that in the SAM method, a brittle elastic behavior of the spandrels that fail in shear, was assumed.

Although the modeling issue is not critical for one, or two-story buildings, it can have a strong impact on the results for buildings with more than two stories.
Further analyses were performed with a five-story wall to assess the impact of several modeling assumptions regarding the strength and stiffness of the connecting elements [perimeter ring beams and masonry spandrels]. The five-story wall (Figure 6) is taken from an existing brick masonry building, with continuous beams on each floor, in the city of Catania, built around 1952. Such a wall was subjected to a "model" of seismic forces, gradually increasing, proportional to a scalar, using different assumptions regarding the connecting elements, as described in Table 1. To cope with a possible softening of the structure, before global collapse is reached, the analyses were carried out by controlling the displacement of a single point of an external, statically determined system, the seismic forces distributed on the floors of which maintain the desired ratio between the forces.
The calculated global resistances (maximum shear force at the base, Vmax) in the different analyses are summarized in Table 1, and the full force-displacement curves are referenced in Figure 7.

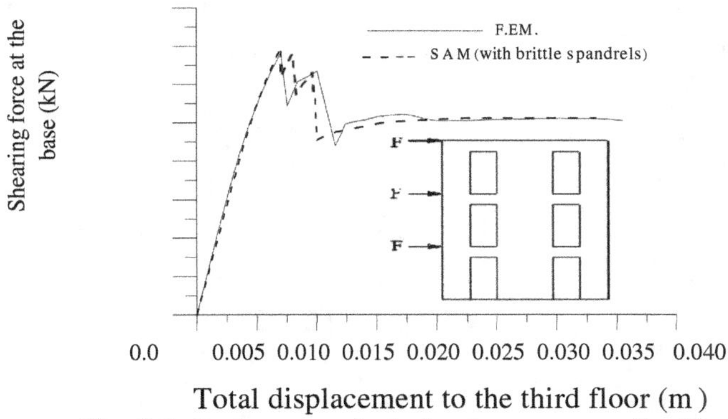

Fig. 8.1- Pushover analysis for three-story wall with weak spandrel

The change in resistance is quite significant, indicating that the influence of the connecting elements on the resistance of a multi-story wall is relatively large, 50% to 100%. At the same time, the global failure mechanism of the wall can vary from a storey mechanism (on the ground floor or the top floor), to a global overturning of the cantilever walls (in case G, where there is no presence of the ring beams), as reflected by the displacement profiles in Figure 8. Such a variety of results shows how the role of the connecting elements should **not be neglected** in a seismic analysis.

The SAM method was also applied to perform a three-dimensional analysis of a five-story building considered in the analysis. The model (approximately 15 x 11 m in plan, 19 m in height) consisted of 390 elements and 195 nodes, for a total of 432 degrees of freedom (assuming the stiffness in the floor plan). As can be seen, such a model can easily be handled by any modern personal computer.

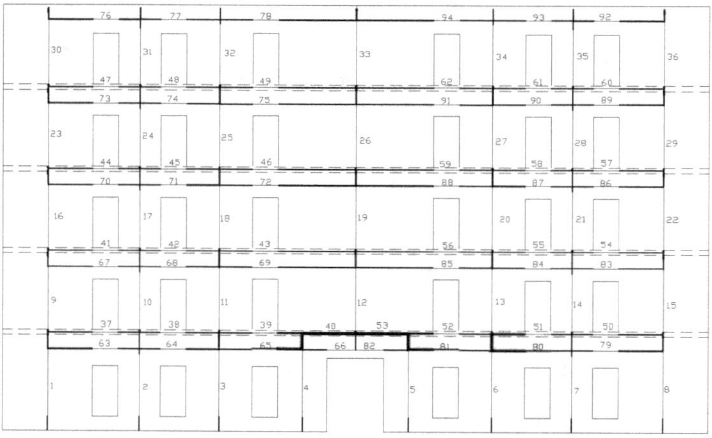

Fig. 8.2. Frame equivalence of a 5-story wall model.

The calculated resistance in the weakest direction, in the most realistic hypotheses, was Vmax/ Wtot = 0.15, which revealed a very high seismic vulnerability. The result is currently being compared with the results achieved by other researchers, with different analytical models.

Tab. 8.1. Summary of the analysis performed on a five-story masonry.

ANALYSIS	HYPOTHESIS	V_{max} [kN]	V_{max}/W_{tot}
A	Ring beams b. a., calculated stiffness for uncracked section	1227	0.369
B	Ring beam b. a., calculated stiffness for cracked section (0.2 of A)	848	0.255
C	Elasto-plastic ring beam; bending stiffness calculated according to possible existing reinforcement	674	0.203
G	Spandrels only, no concrete ring beams	656	0.197
I	Coupling elements, without bending stiffness (cantilever walls)	477	0.143

V_{max} = Max shearing force at the base;
W_{tot} = Total weight of the wall.

Global angular deformation (%)

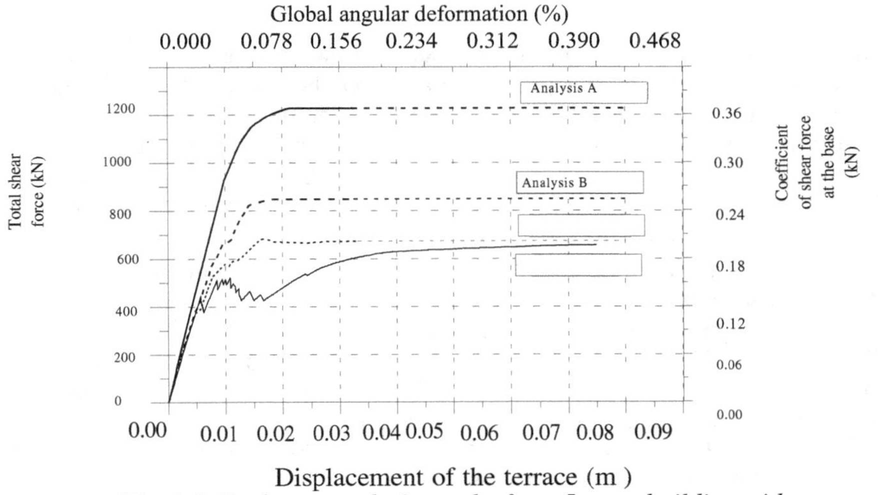

***Fig. 8.3**. Pushover analysis results for a 5-story building with walls.*

8.3-SEISMIC ASSESSMENT PROCEDURE [8]

The following describes the possible use of the proposed model within a simplified seismic assessment procedure. It is assumed that the seismic input is given by the elastic design displacement-acceleration spectrum. The procedure proposed here is based on the use of the displacement spectrum and the "replacement structure" concept [Shibata and Sozen, 1976], which has been adopted in recent proposals for design and assessment according to base displacements [Priestley and Calvi, 1997], which were presented for masonry by Magenes and Calvi [1997] in the case of single-member systems.

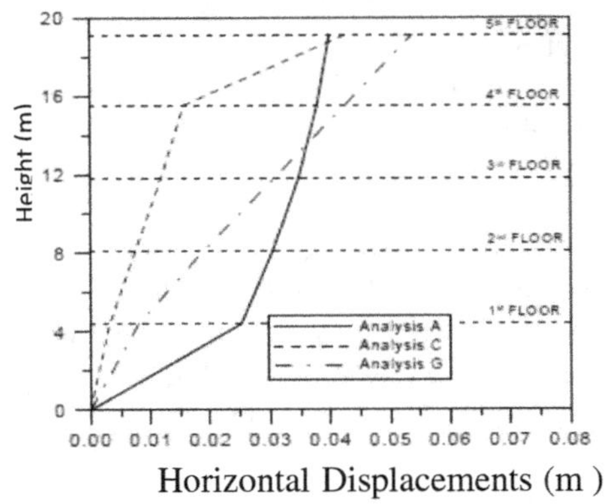

Fig. 8.4. Displacement profiles associated with different collapse mechanisms.

Other approaches could be considered, based for example on the force reduction factor and acceleration spectrum, or based on the composite displacement-acceleration spectrum, and they will be considered in future developments.
The purpose of the procedure is to evaluate the deformed shape of the building, at the peak of the response. First, in this context, it will be assumed that the structure is fairly regular, so the multimodal response will not be taken into account. The main steps of the procedure can be described as follows:

1) We assume a deformed shape $\{\delta(0)\}$, and define the distribution of the equivalent static inertia forces $\{F\}$ as

$$\mathbf{F_i = F_{base} * \gamma_i \quad where \quad \gamma_i = M_i\, d_i^{(0)} / \Sigma\, M_i\, \delta_i^{(0)}} \qquad (8.1)$$

Where $\mathbf{M_i}$ and $\boldsymbol{\delta_i}^{(0}$ are the concentrated mass and horizontal displacement at the "i" degree of freedom, respectively, and $\mathbf{F_{base}}$, the total shear force of the base. The first possible choice, for $\{\boldsymbol{\delta}^{(0)}\}$, can be obtained from the first form related to the initial elastic stiffness of the building, or more simply from a set of displacements that increase linearly with height.

2) A nonlinear static pushover analysis is performed, up to collapse of the structure, under the given static force distribution, maintaining the ratios determined by the coefficients γ_i.
Collapse can be defined as the attainment of the ultimate drift for the individual piers. To address possible structural softening, before reaching the ultimate limit state, it may be necessary to perform static analyses in displacement control, as was done in the examples described in the previous section, to ensure that the desired ratios between seismic forces are maintained.

3) An equivalent system is defined, with the following characteristics:

$$\mathbf{M_{eq} = \Sigma_i\, M_i = M_{tot}} \;; \quad \boldsymbol{\delta_{eq}} = \Sigma \gamma_i \delta_i \qquad \mathbf{F_{eq} = F_{base}} \qquad (8.2)$$

The force-displacement curve $\mathbf{F_{eq}}$-$\boldsymbol{\delta_{eq}}$ of the equivalent system MDF is calculated and plotted. The dynamic response of the system nj.shl will be evaluated after defining a "substitutive structure", the effective stiffness of which will be equal to the shear stiffness $\mathbf{K_{eq,s}}$, at the given value of cdisplacement, $\boldsymbol{\delta_{eq}}$.

4) The equivalent viscous damping ξ_{eq} (including the effects of hysteretic energy absorption) for the replacement structure MDF is determined as an equivalent function of the displacement $\boldsymbol{\delta_{eq}}$, based on the evolution of the damage mechanisms obtained from the pushover analysis and on the principles of energy equivalence. The corresponding curve ξ_{eq}-$\boldsymbol{\delta_{eq}}$ is drawn.

5) The maximum displacement of the system MDF is iteratively estimated, in accordance with the spectrum of elastic design displacements. $\boldsymbol{\delta_{eq,\,max} = SD\ (T_{eq}, \xi_{eq})}$, where:

$$\mathbf{T_{eq} = 2\pi\ (M_{eq}/K_{eq,i})^{1/2}} \qquad (8.3)$$

is the effective period of the response for the maximum displacement. The continuation of steps 1 to 5 is based on the results of the pushover analysis, which is performed using the set of static forces determined in step 1, derived from an accepted deformation mode. However, in the pushover analysis, the displacement ratio of each story may vary due to the nonlinear behavior of the structure and the displacement form corresponding to the value **δeq,**max, calculated at the end of step **5**, will differ from that accepted at the beginning of step **1**.

Depending on the structure, the results of the static analysis may be more or less sensitive to the assumed static force pattern, and in general, the displacement shape will change continuously when the analysis proceeds in the **nonlinear** range, as opposed to the linear shape, or the first mode of vibration. It may therefore be advisable to repeat the procedure, substituting into equation 1, step 1, the displacement shape obtained at the end of step 5, repeating the entire procedure until a final displacement shape is obtained, consistent with the assumed force distribution. However, the need for repetition should not be overestimated.

Giving the pushover approach, it may be more effective to assume two or three **arbitrary** displacement forms, consistent with the most likely failure mechanisms (e.g., floor mechanism in the first floor, floor mechanism in the top floor, global overturning) and then follow steps 1 to 5, once for each accepted displacement form. A range of possible solutions will be obtained, providing a good reference for the assessment. The use of more than one load model is recommended, to account for possible effects of higher modes [Krawinkler and Seneviratna, 1998]. Equations (8.2) in step 3 are obtained from simple principles of dynamic and energy equivalence and do not need special discussion. Step 4 deserves some comments in this context.

The estimation of an equivalent **global** viscous damping for a masonry building requires experimental information on the energy absorption properties of individual structural elements (e.g. piers and spandrels). Once the energy absorption of individual elements has been determined, it is possible to estimate the global energy absorption of the entire structure, and the equivalent global damping. The energy equivalence between the equivalent structure and the building leads to the following expression for the equivalent damping:

$$\xi_{eq} = \mathbf{\Sigma_k\ E_k\ \xi_k / \Sigma_k\ E_k} \qquad (8.4)$$

where $\mathbf{E_k}$ is the elastic strain energy, related to the shear stiffness and $\boldsymbol{\xi}\mathbf{k}$ is the equivalent damping of the **k** structural element. Considering the idealization with equivalent frames of the SAM method, the elastic energy of a single beam-column element can be conventionally expressed in terms of the chord moment and rotation at joints **i** and **j** as:

$$\mathbf{E_k = ½(M_{k,i}\,\theta_{k,i} + M_{k,j}\,\theta_{k,j}) = E_{k,i} + E_{k,j}} \qquad (8.5)$$

where the work due to axial deformation is **neglected**. Currently, limited experimental information is available for masonry structural elements in terms of equivalent damping. References in this document will be made to the work of Magenes and Calvi [1997], where equivalent damping values for brick masonry piers, under static cyclic loading in plan, have been explicitly evaluated.
On this basis, a first rough approximation can be made to determine the amount of viscous damping, equivalent to the hysteretic energy absorption of a single structural element, depending on the mode of failure. In the following application, it is assumed that piers and spandrels, in linear order, are characterized by a constant equivalent damping, equal to **5%**, and that this value increases to **10%** when one of the shear failure criteria is met. If the element fails in bending, the equivalent damping, related to the absorption of hysteretic energy, remains equal to **5%**, but an additional **5%** due to impact and radiation damping is added. The equivalent damping of structural elements will change step by step. These assumptions are intended to give a somewhat **conservative** estimate of the equivalent damping, in relation to the experimental results. The SAM program has therefore been modified to automatically calculate the equivalent damping of the building according to equations (8.4), at each increment of the pushover analysis.

To verify the results that can be obtained from this criterion in a structure, the results of a full-scale cyclic static test on a two-story brick building were processed to obtain a reference for the numerical evaluation of the parameters of the replacement structure. The experiment was carried out at the University of Pavia, [Magenes et al, 1995.] and consisted of a series of displacement cycles, with increasing amplitude, applied to the structure, maintaining a 1:1 ratio between the forces applied to the first and second floors. Walls in the longitudinal direction were connected only by flexible floor beams, so that each longitudinal wall could be analyzed independently, as a two-degree-of-freedom structure. Considering one of the two walls (Figure 9), the experimental response can be evaluated in terms of an equivalent **SDOF** structure, according to the criteria described above, by taking the force–displacement diagram of Figure 10. For each cycle, it is possible to calculate the equivalent damping, based on the absorbed hysteretic energy and the shear stiffness at the peak displacement, by taking the values reported in Fig. 11.

The same wall was also analyzed with the SAM method, performing pushover analysis with equivalent forces at the storey levels, and evaluating the parameters of the replacement SDOF structure, according to the hypotheses described above. Since the test was **static,** impact and radiation damping were not taken into account in the evaluation of **ξeq**. A deformation limit **θu = 0.5%** was assumed for the piers that were destroyed by shear, and the numerical collapse of the structure coincided with the achievement of the limit deformation in the central pier, at the first floor.

8.4 - WALL in axis D – a wall featuring openings.

In general, it can be observed from Figures 8.5 and 8.6 that the analytical method obtains an acceptable estimate of the force-displacement curve, nonlinearity and equivalent damping.

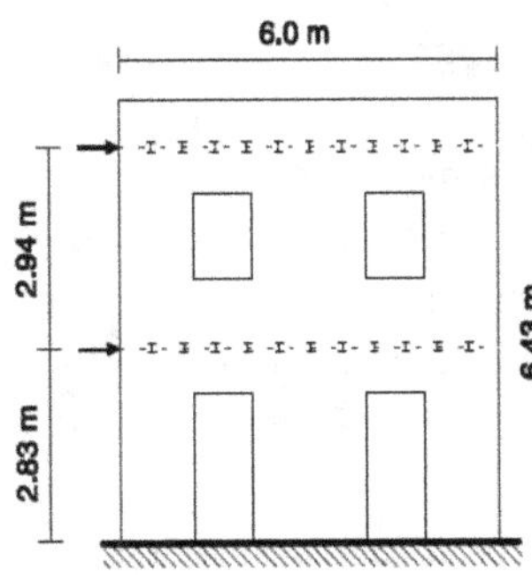

***Fig. 8.5-** Wall on the D axis tested in cyclic load*

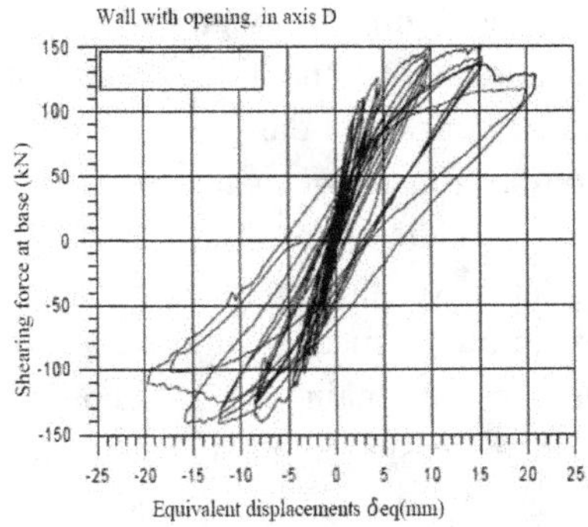

***Fig.8.6-** F-d_{eq} curve (shear force–displacement) calculated from theexperimental reaction of a wall*

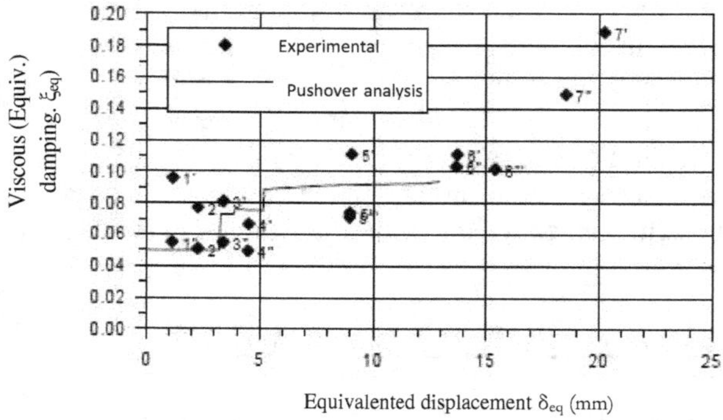

Fig. 8.7- Equivalent viscous damping, related to hysteretic energy absorption.

The experimental cycles beyond the ± 15 mm displacements (labeled 7' and 7") show a significant degradation of the resistance, related to the collapse of the spandrel above one of the doors, which explains the very high experimental value of the equivalent damping. By classifying these cycles as "beyond collapse", the comparison between the experiment and the analysis is meaningful only after cycle 6. The "jumps" in the numerical curve **ξeq** in Fig. 8.8 correspond to the collapse of the pier, or spandrel, in the SAM analysis, according to the assumptions made. A more realistic evolution of the equivalent damping could perhaps be obtained, with a continuous change of the damping of the elements, with increasing angular deformation, although this is not expected to give very different results in the evaluation.

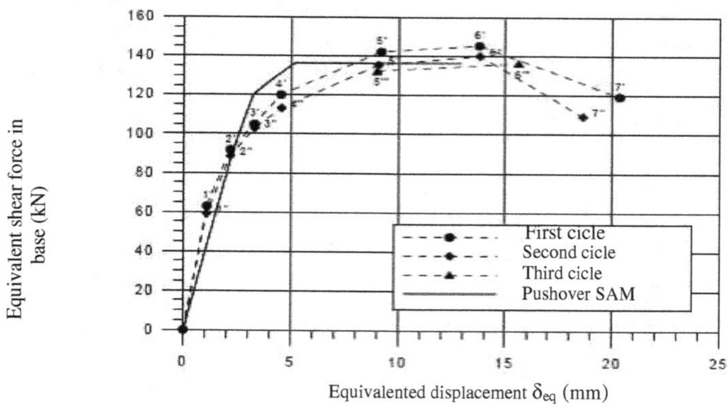

Fig. 8.8- Comparison between experimental envelope and pushover analysis

8.4- A SIMPLE EXAMPLE OF APPLICATION

As an example of the application of the assessment procedure described below, the ideal two-story structure represented by the wall in Figure 9 is assessed, assuming a seismic input described by the elastic response spectrum of Eurocode 8, for the hard soil (soil A). The displacement spectrum is obtained from the acceleration spectrum as:

$$SD(T;\xi) = 4\pi^2/T^2 SA(T;\xi). \qquad (8.6)$$

The elastic spectrum determined for a default extinction of 5% is scaled by multiplying the ordinate by the factor suggested in EC8:

$$\eta = [7/(2+\xi)]^{1/2} \qquad (8.7)$$

A maximum ground acceleration of **0.25g** is assumed.

The distributed masses are concentrated at the floor levels and added to the masses associated with the slabs of each floor, giving a total mass of M1=171.1 kN at the first floor and M2=151.9 kN at the second floor. To perform the pushover analysis, a normalized displacement form, corresponding to the first mode of vibration, is assumed in step 1:

$$\{\delta^{(0)}\}^T = \{0.545\ ;\ 1.0\} \qquad (8.8)$$

which gives a distribution of forces:

$$\{F^{(0)}\}^T = F_{base}\{0.38\ ;\ 0.62\} \qquad (8.9)$$

The pushover analysis is then performed (Step 2), and the F_{base} - δ_{eq} and ξ_{eq}- δ_{eq} curves of a one degree of freedom structure are obtained (Steps 3 and 4). Step 5 is then performed by assuming the first value of $\boldsymbol{\delta_{eq,0}}$ equal to the finite displacement of the $\mathbf{F_{base}}$ - $\boldsymbol{\delta_{eq}}$ curve. The corresponding shear stiffness $\mathbf{K_{eq}=F_{eq}/\delta_{eq}}$, the period $\mathbf{T_{eq}}$ and the damping $\boldsymbol{\xi_{eq}}$ are evaluated and the displacement spectrum is inserted to obtain a new displacement value, $\boldsymbol{\delta_{eq,1}}$. In the first approximation, by checking whether $\delta eq_{,1} \leq \delta_{eq,0}$, it is possible to verify whether the final displacement $\delta_{eq,ult}$, of the structure will not be exceeded. If this is verified, by a trial and error procedure, it is possible to converge to a final value of the displacement, such that:

$$\boldsymbol{\delta eq,n+1 \cong \delta eq,n,}$$

within a certain tolerance. In the considered case this results in $\delta_{eq,max}$ =12.1mm, compared to a final settlement $\delta_{eq,ult}$ = 13.6 mm. At this point, the displacement

mode $\{\delta^{(1)}\}$ corresponding to the "design" displacement of 12.1 mm can be checked and compared to the initial accepted displacement mode $\{\delta^{(0)}\}$.
In this case, after normalization, the displacement form $\{\delta^{(1)}\}^T = \{0.804; 1.0\}$ is obtained, which indicates a floor-to-floor mechanism. As can be seen, the calculated inelastic response can lead to a displacement form that is quite different from the first elastic modal form.
If now a new distribution of seismic forces $\{F^{(1)}\}$ is calculated from $\{\delta^{(1)}\}$, taking into account:

$$\{F^{(1)}\}^T = F_{base}\,\{0{,}475;\ 0{,}525\}$$

steps 1 to 5 can be repeated, defining new curves:

$$\mathbf{Fbase - \delta_{eq}\ dhe\ \xi_{eq} - \delta_{eq}} \qquad (8.9)$$

It may be worth noting that now the force distribution approximates a constant force distribution, which is consistent with the obtained floor mechanism. At the end of this global approximation the following results are obtained::

$$\delta eq,\ max = 5.5\ mm,\ \{\delta\ (2)\}\ T = \{0{,}668;\ 1.0\}.$$

Further approximation gives: $\delta_{eq,\ max} = 6.0$ mm, $\{\delta^{(3)}\}^T = \{0{,}665;\ 1.0\}$.

The similarity of the values calculated in the last two approximations suggests that the δeq value, max = 6.0 mm can be considered as an acceptable estimate of the maximum response of the structure.

CHAPTER 9

ANALYSIS OF A 5-FLOOR BUILDING, TYPE 77/5, WITH LOAD-BEARING BRICK MASONRY

Fig. 9.1 3D view of building 77/ 5

9.1. SOME CONSIDERATIONS for building type 77/5

The reason this building was chosen for analysis is primarily due to its external form, the L-shaped floor plan it has, where a pronounced scaling is evident at the entrance, making it more vulnerable to seismic reactions.
Below we list some of the main reasons that led us to this choice:

1- The closed-quadrant organization of the building has not been fully respected,
2- The beams necessary for absorbing seismic forces in some axes are missing, and so are the seismic columns.
3- The E and F axes are offset from each other by 70 cm, creating the possibility of a rotational moment.
4- There is a lack of anti-seismic columns in its corners, especially in the outer walls.

To illustrate these proposals, the following are some sheets from the project of this building, which were taken from the Central Construction Archive. The recommendations of the foreign and domestic literature on which this analysis is based are given below. The following is also a computer analysis with:
a- The ETABS program, which highlights the lack of stability, periods and displacements greater than those listed.
b- The AM-quake program, which highlights the steps after which the building collapses.
At the end of the study, the respective suggestions are given to improve the deficiencies identified above.

9.2 DATA DISCRETIZATION

The 5-storey building, type 77/5, with a structure of load-bearing walls, described above was discretized to be analyzed with the ETAB program.
The ground and first floors of this the building is made of brick masonry thick t=38cm, while the upper floors are made of brick masonry with a thickness of t=25cm. At the top of the masonry of each floor, following the slab, are concrete strips (38 x15) cm and (25x15) cm, depending on the thickness of the walls. The parapet is made of brick walls with a thickness of t=12cm and a height of h=80cm. The floors in this building are slabs with beams, with a width of b=8 cm and a height of h=13cm, with an interaxle of 28 cm. So, we have a mixture between prefabrication and monolithic. (see the attached drawings 9.1-9.6)

9.3 BRIEF ARCHITECTURAL AND CONSTRUCTION PLANS

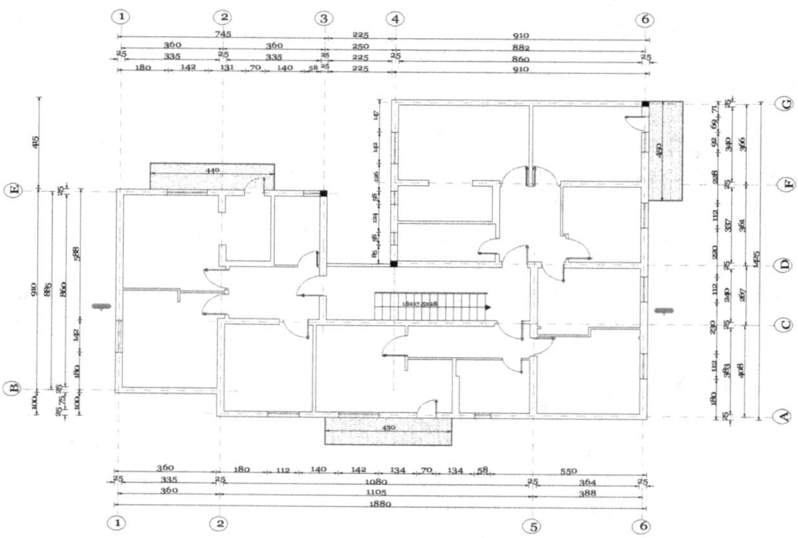

Fig 9.1- TYPICAL FLOOR PLAN

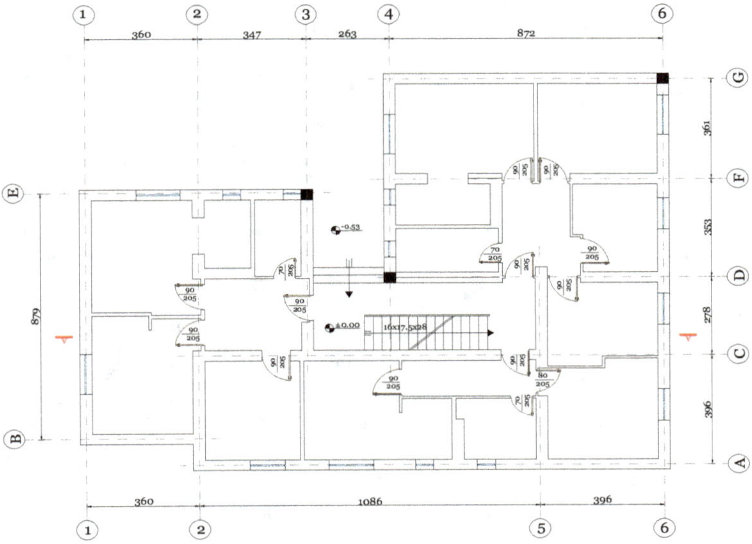

Fig 9.2- GROUND FLOOR PLAN

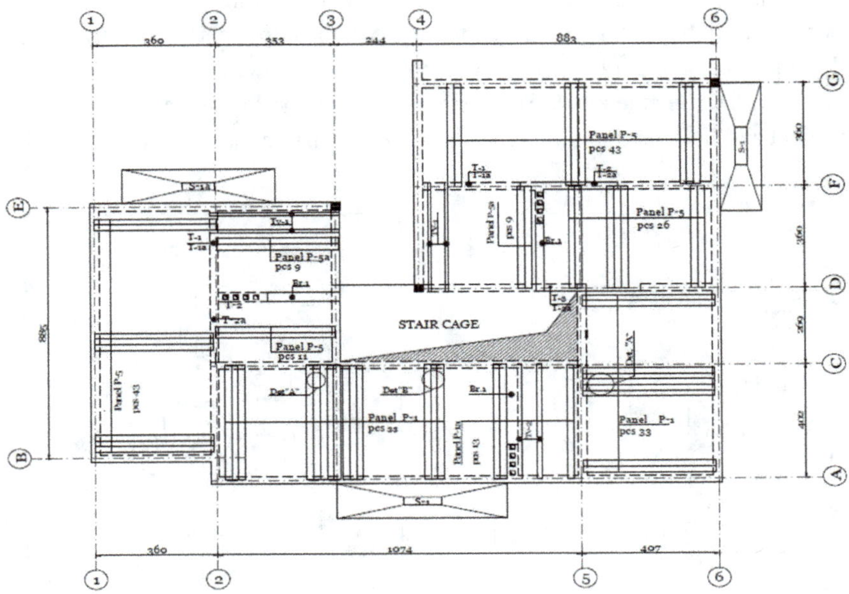

Fig 9.3 - PLAN OF TERRACE STRUCTURES

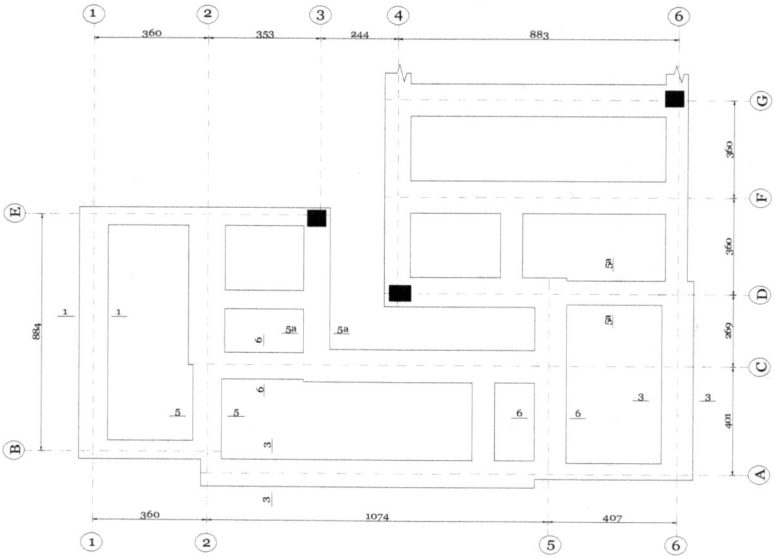

Fig 9.4 - FOUNDATION PLAN

The beams served simply as formwork; no concrete filling was foreseen in the upper area as in beam slabs today. In these slabs, the upper part of the brick beam had some recesses, about 4cm x 4cm, in which iron wires were placed, with cement mortar. (See the Fig. 9.6). This results in the horizontal diaphragm lacking sufficient rigidity, making it more susceptible to seismic movement.

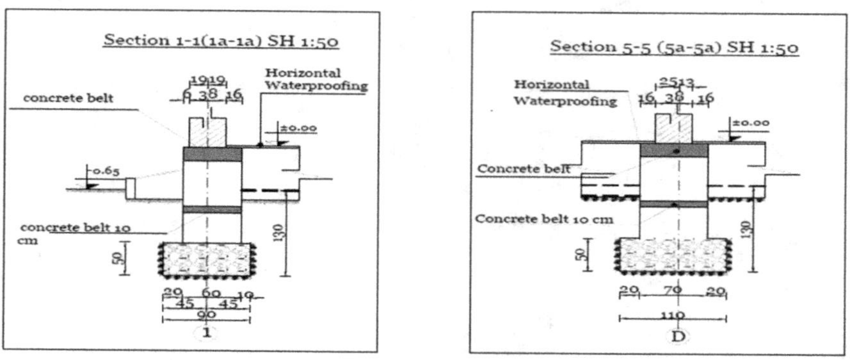

Fig 9.5- FOUNDATION SECTION

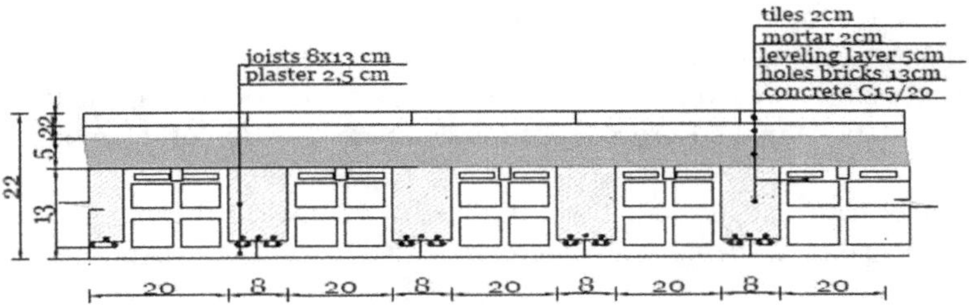

Fig 9.6- SLAB SECTION

9.4 ANALYSIS OF ACTIVE LOADS on the BUILDING

Loads per beam 28 cm width (8 +10+10) :

1- Cement tiles	0.02 x 2200 x 0.28	= 12.30 kg/ml
2- Leveling layer	$0.06 \times 1800 \times 0.28$	$= 30.2 kg/$ml
3- Cement mortar substr.	$0.03 \times 1800 \times 0.28$	$= 15.1 kg/$ml
4- Hole bricks	$0.20 \times 0.13 \times 800$	$= 20.8 kg/$ml
5- Plastering 2.5 cm	$0.025 \times 0.28 \times 1800$	$= 12.6 kg$ /ml
	Shuma:	**91.10 kg/ml**

For 1 m2 we have: $\frac{100}{28} 91.10 = 325.4\ kg/$m2 = 3.25 kN/m2

Finally:

- Permanent load (layers and slabs): D= 3.25 kN/m^2
- Temporary load: L = 2.0 kN/m^2

The weight of the partition walls, with a no-hole brick, the thickness of t=12 (taking into account the weight of the 2.5 cm plaster on both sides) it is accepted as 280 kg/m2 (2.8 kN/m2).

The quality of concrete has not met today's expectations. To accurately reflect the concrete quality of that period, for this study purpose, we accepted the C15/20 brand concrete as input for the program.

The brick used in the masonry of these buildings is red clay brick, measuring 6 x 12 x 25 cm, and has a strength of f_k = 7.5 N/mm2

According to the sketch in Fig. 9.6, in the ETABS program, the beams are designed as C15/20 secondary beams, measuring 8 cm by 13 cm, are spaced every 20 cm apart. The design excludes the possibility of a slab on the upper beam. The loads from the sub-layers, formwork bricks, and partition walls are transferred to these beams. Also, a 2 cm thick concrete leveling layer is assumed, which correctly approximates the real situation of the floor.

The coefficient of elasticity for brick masonry material, following the recommendations of Eurocode 6, is accepted as **E=1000** N/mm2 for service

conditions. For the ultimate limit state, primarily in nonlinear analyses, it is recommended to use a value of **E=600** N/mm2.
Various comparisons with experimental values (Tomazevic, 1999) indicate that: 'the recommendations in the Eurocode lead to an overestimation of the modulus of elasticity.'
Thomas Zimmermann and his co-authors (Zimmermann et al., 2012) recommend the following equation as being more appropriate and closer to the experimental values.

$$E=300 \quad f_k \quad [N/mm^2]$$ [5]

(8.10)

The determination of the compressive strength value according to KTP-9-78 also dependent on the strength of the mortar and that of the masonry (“mortar grade” and “brick grade”).

TAB 9.6 - Strength value f_k for different strength values of brick-and-mortar combinations

Nr	Brick brand	Mortar brand (N/mm2)						
	N/mm2	10	7.5	5.0	2.5	1.5	0.4	0.0
1	15	2.2	2	1.8	1.5	1.35	1.2	0.8
2	10	1.8	1.7	1.5	1.3	1.1	0.9	0.6
3	7.5	1.5	1.4	1.3	1.1	0.9	0.7	0.5
4	5.0		1.1	1.0	0.9	0.75	0.6	0.35

Based on the accepted values, the brick = **7.5** N/mm2 and the mortar= **2.5** N/mm2, the **Table 9.6** gives the following resistance of the masonry:

$$f_k = 1.1 \text{ N/mm2}$$

According to the author Thomas Zimmermann for the modulus of elasticity recommendations, we accept the following value:

$$E=300 \times f_k = 300 \times 1.1 = 330 N/mm^2$$

9.5 CENTER OF MASS DETERMINATION for the standard floor

Static moments for each wall in the directions **X**:

Tab. 9.1

Walls in direction Y				
L(m)	**Yc**	**T(m)**	**Sx**	**Fy**
5.5	6.97	0.38	14.5673	2.09
1.42	2.1	0.38	1.13316	0.5396
0.63	9.4	0.38	2.25036	0.2394
0.52	7.53	0.38	1.487928	0.1976
4.33	2.16	0.38	3.554064	1.6454
2.83	8.3	0.38	8.92582	1.0754
1.0	4.83	0.38	1.8354	0.38
1.09	13.32	0.38	5.517144	0.4142
1.26	10.73	0.38	5.137524	0.4788
1.3	8.88	0.38	4.38672	0.494
0.54	7.38	0.38	1.514376	0.2052
1.48	1.12	0.38	0.629888	0.5624
1.6	9.48	0.38	5.76384	0.608
3.62	5.48	0.38	7.538288	1.3756
2.3	1.53	0.38	1.33722	0.874
0.41	13.66	0.38	2.128228	0.1558
2.28	10.9	0.38	9.44376	0.8664
2.3	7.49	0.38	6.54626	0.874
2.3	4.07	0.38	3.55718	0.874
1.8	0.9	0.38	0.6156	0.684
	Sum		**87.87006**	**14.6338**

Walls in direction X				
L(m)	**Yc**	**T(m)**	**Sx**	**Fx**
11.04	0.19	0.38	0.797088	4.1952
3.98	1.19	0.38	1.799756	1.5124
8.74	4.14	0.38	13.74977	3.3212
1.98	7.08	0.38	5.326992	0.7524
6.6	6.73	0.38	16.87884	2.508
4.87	9.91	0.38	18.33945	1.8506
4.88	10.47	0.38	19.41557	1.8544
9.1	14.06	0.38	48.61948	3.458
	Sum		**124.9269**	**19.4522**

$$s_x = s_{x(y)} + s_{x(x)} = 87.87 + 124.93 = 212.80$$
$$F = F_x + F_y = 14.634 + 19.452 = 34.086m^2$$

Static moments for each wall in the directions **Y:**

Tab. 9.2

Walls in direction X			
Lm	**Xc**	**tm**	**Sy**
1.42	4.69	0.38	2.530724
1.4	7.22	0.38	3.84104
1.33	10	0.38	5.054
1.4	11.95	0.38	6.3574
5.19	15.82	0.38	31.2002
3.6	1.8	0.38	2.4624
1.82	4.89	0.38	3.381924
6.72	10.06	0.38	25.68922
0.2	14.42	0.38	1.09592
1.98	4.6	0.38	3.46104
3.78	11.6	0.38	16.66224
2.45	17.2	0.38	16.0132
1.8	0.9	0.38	0.6156
1.35	3.9	0.38	2.0007
1.33	5.81	0.38	2.936374
0.38	7.26	0.38	1.048344
1.02	10.6	0.38	4.10856
1.08	12.94	0.38	5.310576
2.78	17.02	0.38	17.97993
9.1	14.25	0.38	49.2765
Shuma			**201.0259**

Walls in direction Y			
Lm	**Xc**	**tm**	**Sy**
6.92	0.19	0.38	0.499624
5.86	3.79	0.38	8.439572
3.83	7.26	0.38	10.5662
4.18	9.9	0.38	15.72516
1.48	12.04	0.38	6.771296
1.6	13.4	0.38	8.1472
5.92	14.71	0.38	33.09162
9.09	18.42	0.38	63.62636
Shuma			**146.867**

$$s_y = s_{y(x)} + s_{y(y)} = 201.03 + 146.87 = 347.90$$

Now we can determine the center of rigidity coordinates:

$$\boldsymbol{y_c = \frac{s_x}{F} = \frac{212.80}{34.086} = 6.24m} \quad \boldsymbol{x_c = \frac{s_y}{F}} = \boldsymbol{\frac{347.90}{34.086}} = 10.21m$$

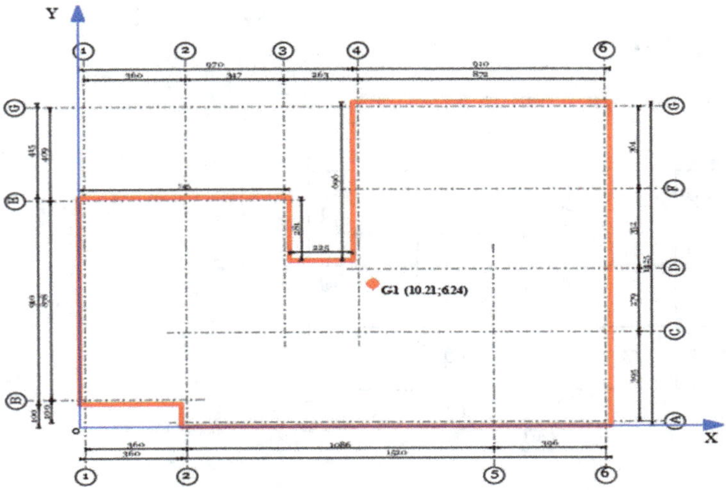

Fig. 9.7- Mass Center of the typical floor

GEOMETRIC ANALYSIS OF THE TYPICAL FLOOR PLAN

[Center of gravity determination]

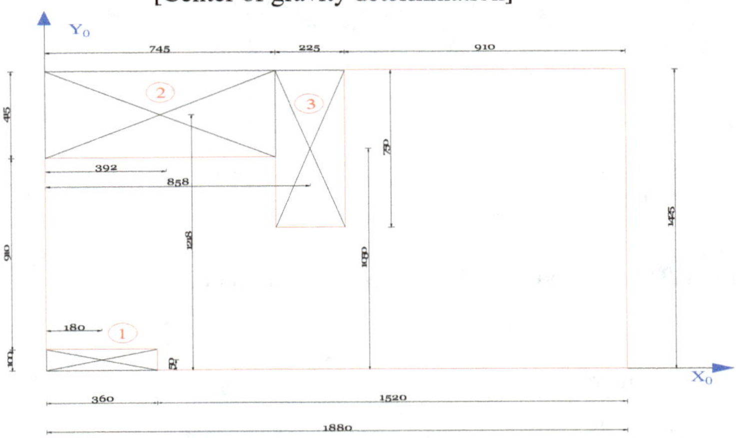

Fig. 9.8- Determination of the center of gravity of a typical floor

Tab. 9.3 Static moments Sy

a	b	A	xc	Sy
18.8	14.25	267.9	9.4	2518.26
3.6	1	3.6	1.8	6.48
7.45	4.15	30.918	3.725	115.1677
2.25	7.5	16.875	8.58	144.7875
Shuma		216.51		2251.825

$X_C = Sy/A = 2251.82/216.51 = 10.40$

Tab. 9.4 Static moments Sx

a	b	A	yc	Sx
18.8	14.25	267.9	7.125	1908.788
3.6	1	3.6	0.5	1.8
7.45	4.15	30.918	12.21	377.5027
2.25	7.5	16.875	10.48	176.85
		216.51	**Sum**	**1352.63**

$Y_C = Sx/A = 1352.63/216.51 = 6.25$

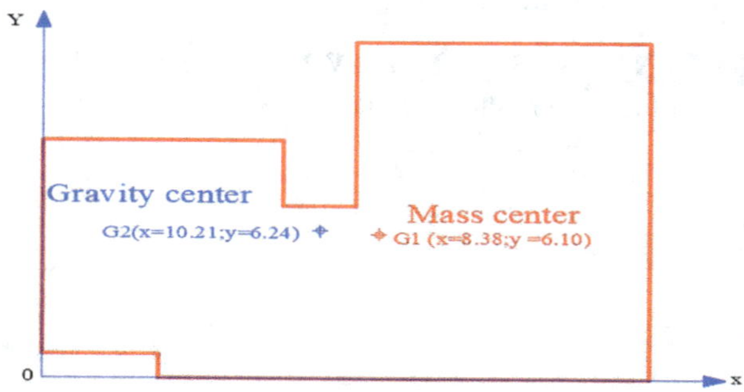

Fig.9.9- Mass and Gravity centers, their coordinates.

Y_C/ Xc	Gravity center(m)	Mass center (m)
Y_C	6.24	6.25
X_C	10.21	10.4

9.6.1- DETERMINATION of inertia and resistance moments relative to the CENTER OF GRAVITY

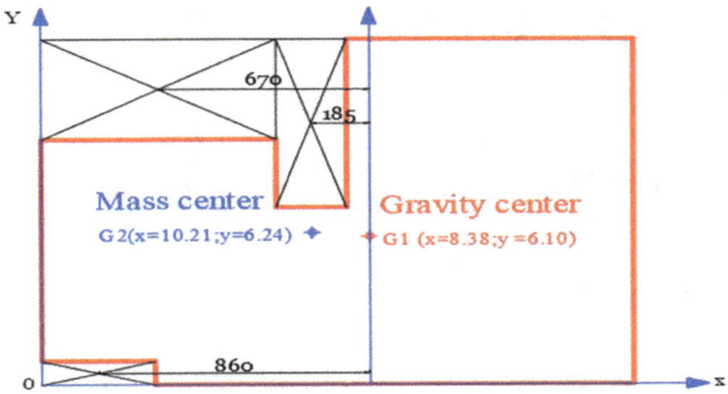

Fig. 9.10 - Distances from the axis Y_0-Y_0 of missing parts

a) Moment of inertia I-x about the axis $X_0 - X_0$

$$I_X^0 = \frac{bh^3}{12} = \frac{(15.2 + 3.6) \cdot 14.25^3}{12} + (18.8 \cdot 14.25) \cdot 0.87^2 = 4533 + 202.8 = 4736\ m^4$$

Let's deal with the empties in the plan:

1. $$\frac{3.6 \cdot 1^3}{12} + 3.6 \cdot 1 \cdot (6.25 - 0.5)^2 = 119.32\ m^4$$

2. $$\frac{7.45 \cdot 4.15^3}{12} + (7.45 \cdot 4.15) \cdot (14.25 - 6.25 - 2.07)^2 = 44.37 + 1087 = 1131\ m^4$$

3. $$\frac{2.25 \cdot 7.5^3}{12} + (2.25 \cdot 7.5) \cdot (14.25 - 6.25 - 3.75)^2 = 79.1 + 304.8 = 384\ m^4$$

$$\sum I_{X_0} = 4736 - 119 - 1131 - 384 = 3102\ m^4$$

Resistance moments about the axis Y

$$W_x^1 = \frac{3102}{6.24} = 497\ m^3 \qquad W_x^2 = \frac{3102}{(14.25 - 6.24)} = 388\ m^3$$

b) The moment of inertia about the axis Y0 when it coincides with the center of gravity

$$I_Y^0 = \frac{bh^3}{12} = \frac{14.25 \cdot (15.2 + 3.6)^3}{12} = 7890\, m^4$$

Let's deal again with the empties in the plan:

1. $\frac{1 \cdot 3.6^3}{12} + (3.6 \cdot 1) \cdot (8.6)^2 = 3.88 + 266.2 = 270.1\, m^4$

2. $\frac{4.15 \cdot 7.45^3}{12} + (7.45 \cdot 4.15) \cdot (6.68)^2 = 143 + 1379.6 = 1522.61\, m^4$

3. $\frac{7.5 \cdot 2.25^3}{12} + (7.5 \cdot 2.25) \cdot (1.83)^2 = 7.11 + 56.51 = 63.62\, m^4$

$$\sum I_{Y_0} = 7890 - 270.5 - 1522.61 - 63.6 = 6303.8\, m^4$$

The moment of resistance about the axis Y_0 when it coincides with the center of gravity:

$W1y0 = \frac{6303}{8.40} = 750.45\, m^3$ $\quad$ $W2y0 = \frac{6303}{10.40} = 606.05\, m^3$

9.6.2 BENDING MOMENTS at the foundation due to the difference between the center of mass and the center of gravity

Tab. **9.5** Dead load of the building:

Floor	Wall load	Progressive walls Load	Slab load	Progressive slab Load	Total Load
5	142.5	142.5	185	185	327.5
4	142.5	285	185	370	655
3	142.5	427.5	185	555	982.5
2	217.5	645	185	740	1385
1	217.5	862.5	185	925	1787.5

Total Load on the ground floor = 1787.5 ton

Analysis in the direction X-X

The moment due to eccentricity of the loads is:

$$M_Y = 1787.5 \cdot (10.40 - 10.21) = 339.62\ ton \cdot m$$

$$\sigma_1 = \frac{339.6 \cdot 10^5}{750 \cdot 10^6} = \frac{340}{7500} = 0.045\ kg/cm^2$$

$$\sigma_2 = \frac{339 \cdot 10^5}{606 \cdot 10^6} = \frac{339}{6060} = 0.056\ kg/cm^2$$

The influence of the mismatch between the center of mass and the center of gravity in this direction is minimal, as the difference between the two centers is only 19 cm.
The moments in the Y-Y direction are zero due to the ideal alignment of the centers of gravity and mass in this direction

9.7 SEISMIC ANALYSIS

- Seismic hazard parameters. Determination of the response spectrum type:

The seismic risk for the city of Tirana is considerable. Based on some previous studies, (Sulstarova and Aliaj, 2001), (Muco, 2004) the value of the seismic acceleration is accepted as ag=0.2g, as the seismic acceleration ag for Type C soil. To assess the object under the most unfavorable seismic conditions, we refer to the map of Albania's seismogenic zones, where the maximum magnitude is noted as Ms = 5.5-6.5. The periods *TB, TC, TD*, and the soil factor S determine the shape of the response spectrum. For an object located in an area with Ms > 5.5, we select Type 1 of the response spectrum.

$0 \leq T < T_B$ $$S_d(T) = a_g\, S\left[\frac{2}{3} + \frac{T}{T_B}\left(\frac{2.5}{q} - \frac{2}{3}\right)\right] \quad (8.11)$$

$T_B \leq T < T_C$ $$S_d(T) = a_g\, S\, \frac{2.5}{q} \quad (8.12)$$

$T_C \leq T < T_D$ $$S_d(T) = \begin{cases} a_g\, S\, \frac{2.5}{q}\left[\frac{T_C}{T}\right] \\ \geq \beta\, a_g \end{cases} \quad (8.13)$$

$T \leq T_D$ $$S_d(T) = \begin{cases} a_g\, S\, \frac{2.5}{q}\left[\frac{T_C\, T_D}{T^2}\right] \\ \geq \beta\, a_g \end{cases} \quad (8.14)$$

where :
T - the vibration period of a linear system with one degree of freedom;
a_g - The design acceleration of the ground in rock or a rock-like formation. For type B ground, ag = $\gamma_I a_{gR}$, where γ_I is the importance factor and a_{gR} is the maximum value of the ground acceleration referred to the design earthquake for a return period of 475 years.
S - is the soil factor;

T_B. T_C. T_D - are values of the periods, defined in the relevant tables of the Eurocode 8;
β - is the lower limit factor that is recommended to have the value 0.2;

Type "1" of the response spectrum refers to an earthquake with magnitude $M_S >$ 5,5;
Type "2" of the response spectrum refers to an earthquake with magnitude $M_S \leq$ 5,5).
In design practice, using a spectrum envelope can be effective; this spectrum is created to display, for a given period T, the largest spectral value obtained from the comparison of Type 1 and Type 2 spectra.

9.7.2 - DETERMINATION OF THE BEHAVIOR FACTOR

According to Eurocode 8, the upper limit value of the behavior factor "**q**", used to calculate the energy absorption that occurs in critical zones of plastic joints, formed during strong earthquakes, will be determined as follows:

$$q = q_O \cdot k_w \geq 1.5$$

For a regular system, class M (average DC-M) ductility, the q_0 base value calculated is:

$$q_O = 3 \cdot (\alpha_u / \alpha_1)$$

and depends on the type of structural system and the regularity in height. For design purposes based on Table 5.9.1, Pg. 373 (Seismic Engineering, Niko Pojani), for masonry systems, medium ductility, the value of q_0 is

$$q_0 = 3$$

k_w- factor reflecting the prevailing mode of possible destruction, equal to:

$$k_w = (1 + \alpha_O)/3 \leq 1$$

For frame-type systems and dual systems that are equivalent to frame systems $k_w = 1,0$;
In structural masonry systems, the value is $k_w = 0.5$.
From the above values we determine the behavior factor value:

$$q = q_0 \times k_w = 3 \times 0.5 = 1.5$$

It should be noted that, in principle, the values of **q** are adjusted, considering the designer's choice for the given deformation of the structure, i.e., the choice of the ductility class. Below is a table from the ETABS program that shows the type of spectrum and other seismic data for the studied object

9.7.3. Soil CATEGORY

For the soil category, we chose the most unfavorable case, **Category C**, which consists of deep deposits of compacted or semi-compacted sand, gravel, or hard clay, with thicknesses ranging from several tens to several hundred meters.

9.7.4- DESIGN RESISTANCE VALUES for materials

The brick material data entered in ETABS is:

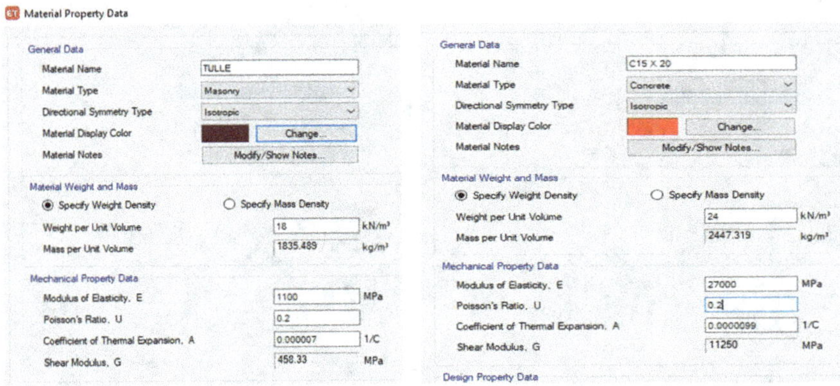

Fig. 9.11 – The brick and concrete parameters, input into Etabs

9.7.5 - Combination of the effects of the seismic action components. [6]

Recommendations of **EN-1998-1**, p. 65, Combination of seismic action components effects in the light of:
P (1) In general, horizontal components of seismic action (see **3.2.2.1(3)**) shall be considered as acting simultaneously.
(3) As an alternative, the effects of action, due to the horizontal components combination of seismic action, may be calculated using the following combinations:

a) E_{Edx} "+ " $0.3\,E_{Edy}$ (4.18)

b) $0.3E_{Edx}$ "+ " E_{Edy} (4.19)

where: "+" means " to combine with"
E_{Edx} - represents the effects of seismic action applied along the structure's horizontal axis x;
E_{Edy} - represents the effects of seismic action applied along the structure's horizontal y-axis Based on these recommendations, these combinations have been used in the analysis in ETABs:

Comb RSX = Dead + 0.3 Live+ RSX + 0.3 RSY
Comb RSY = Dead + 0.3 Live+ 0.3 RSX + RSY
Comb ELX = Dead + 0.3 Live+ EQLX +0.3EQLY
Comb ELY = Dead + 0.3 Live+ 0.3EQLX + EQLY

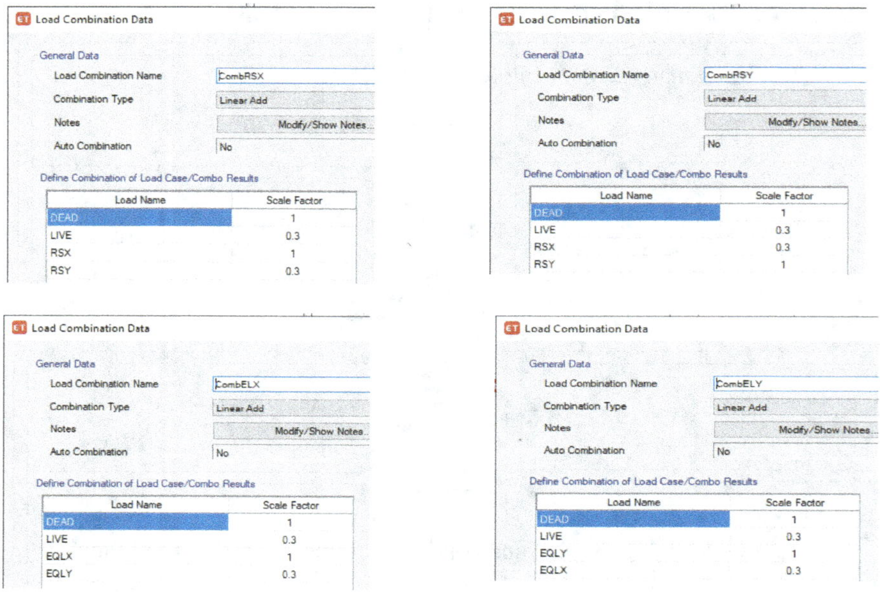

Fig. 9.12 – Table 1 of Etabs combinations

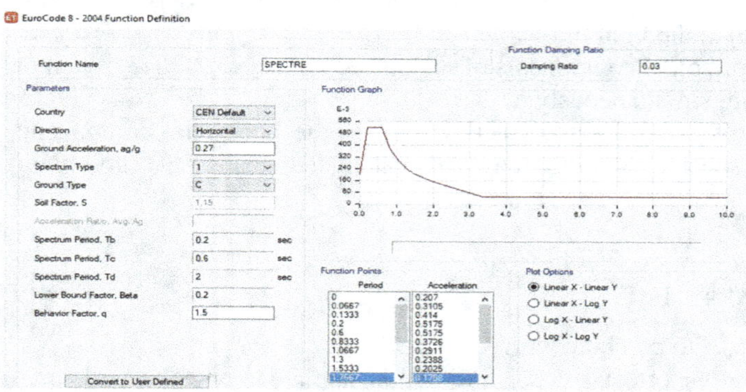

Fig. 9.13 - Response spectrum, seismic acceleration, soil category.

The **RSX component** is derived from the response spectrum selected for the calculations, and the same applies to RSY.

Response spectrum chosen is **Eurocode 8-2004**

3.8 THE RESULTS

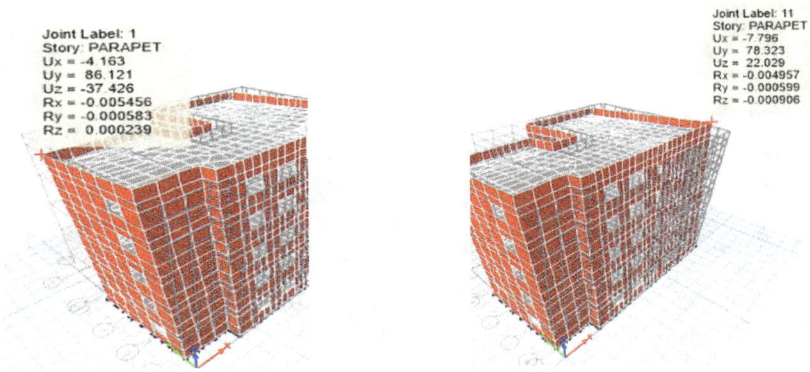

Fig. 9.14- Maximum displacements at the building top

9.8.1 - DISPLACEMENTS:

The building **permitted** horizontal displacements are 1/200 H = 7.1 cm
The **maximum** displacements for the elastic phase are = **8.1cm > 7.1 cm** (for comb. ELY):

D+ 0.3L+ 0.3EQLX+ EQLY

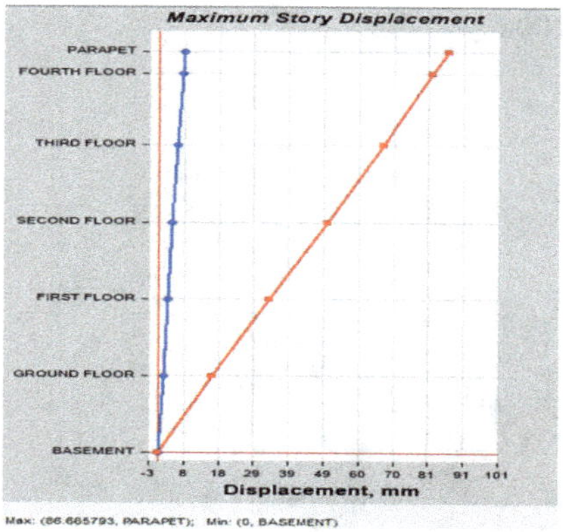

Fig. 9.15 X and **Y** displacements (**Uy** = 86.1 mm)

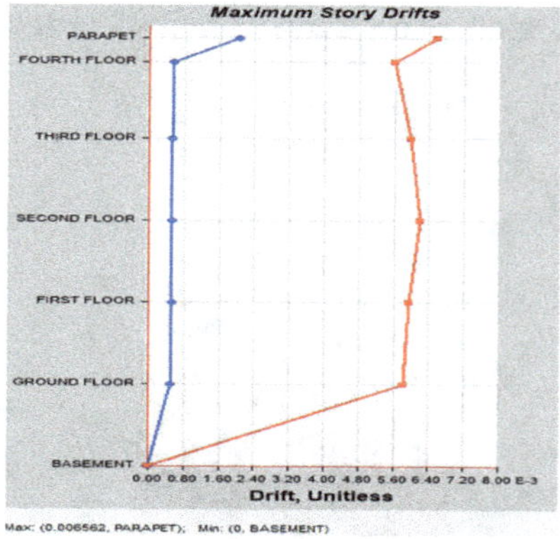

Fig. 9.16 Drifts according to X and Y (Combination EQLY)

The maximum displacements for the plastic phase are:

8.1cm x q = 8.1X 1.5= 12.15 cm

9.8.2- THE PERIOD

According to EC8 recommendations, for masonry buildings, the period of natural oscillations should be:

$$[T] = 0.05 \text{ x } Hg^{0.75} = 0.05\text{x } 14.2^{0.75} = [0.366] << T = 0.690 \text{ s}$$

Where **0.690 s** is the period of the building in the analysis, in the first form of oscillation.

Tab. 9.7 (7.8.III) – *Recommended area of resistant masonry in each orthogonal direction*

for simple constructions **[2]**

Acceleration on the top of the terrain ag·S		≤0,07 g	≤0,1g	≤0,15 g	≤0,20 g	≤0,25 g	≤0,30 g	≤0,35 g	≤0,40 g	≤0,45g	≤0,47 g
Type of structure	Number of floors										
Conventional masonry	1	3,5 %	3,5 %	4,0 %	4,5 %	5,0 %	5,5 %	6,0 %	6,0 %	6,0 %	6,5 %
	2	4,0 %	4,0 %	4,5 %	5,0 %	5,5 %	6,0 %	6,5 %	6,5 %	6,5 %	7,0 %
	3	4,5 %	4,5 %	5,0 %	5,5 %	6,0 %	6,5 %	7,0 %			

According to the recommendations in the table above for a **3-story** building and an acceleration of **ag=0.25g** (the table does not include buildings higher than 3 floors), the percentage of masonry is **6%** in each direction.
While, for the 77/5 building, the masonry percentages are **18%** and **17%** in the X and Y directions, respectively. Despite this high percentage of masonry, the building exhibits spans twice as large as allowed, which shows that it is **paradoxically** flexible..
It is evident that the large weight of retaining walls, whose presence is evidenced in the high percentage of walls, has a significant impact on these parameters of the period.

Therefore, if we consider the same percentage of masonry, while altering its weight (in a hypothetical scenario), a significant decrease in the period is observed, as demonstrated below. For normal-weight brick walls with a density of 18 kN/m³, we have the following period:

- Mode 1 - Period 0.656276594155882

While for wall weight 10 kN/m3, (hypothetical scenario) this oscillation period will be:

Mode 1 - Period 0.51674233117687

Thus, we will experience **a reduction of 16.5%** in the first mode.
However, this is merely an attempt—a hypothesis aimed at identifying the causes and potential solutions for the shortcomings revealed by the analysis. The period of the building under analysis is far from what is recommended by the standards.

During several years of researching the technical literature on masonry buildings, I have not encountered any recommendations regarding the maximum percentage of masonry for buildings taller than three floors, aside from the table above. I think the main reason is that various authors have excluded the possibility of constructing buildings with masonry, higher than 10m in seismic zones.
Based on observations from the November 26, 2019 earthquake in the Durres region, these types of buildings responded relatively well, especially when no reduction interventions were made to the masonry on their ground floors.
This conclusion does not exclude immediate collapse from seismic forces, as long as the building exhibits significant structural deficiencies, the main one being the high period, almost twice as high as those allowed..

9.8.3- <u>LOCAL CONTROLS</u>

The wall on the axis **<u>2-2</u>**

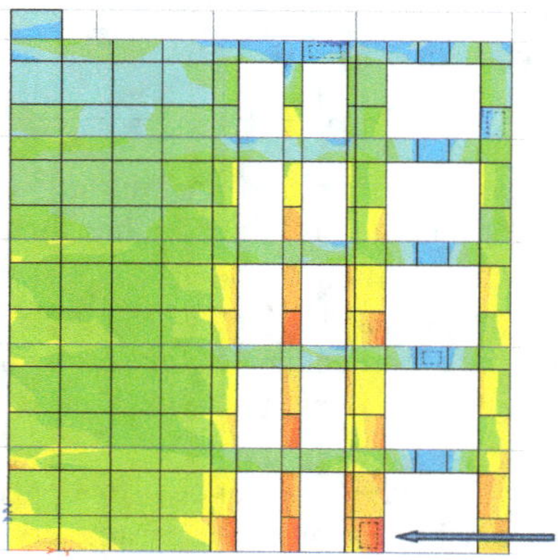

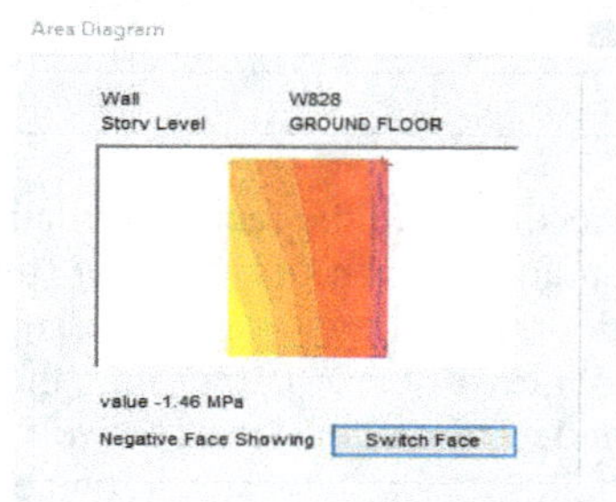

The most stressed element **W 828**

Fig. 9.18 Stresses S2-2 in the element W828

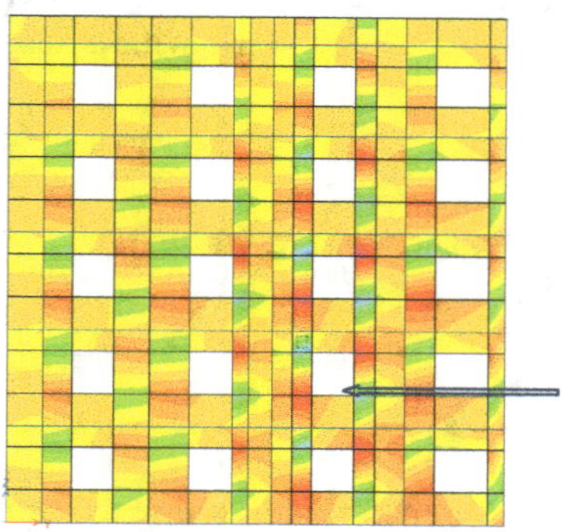

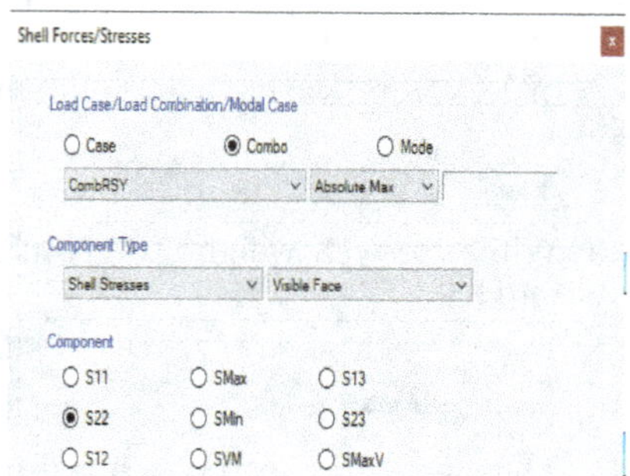

The maximum value on this axis is the S2-2 pressure, with the value: **-1.46 MPa > - 1.1MPa**

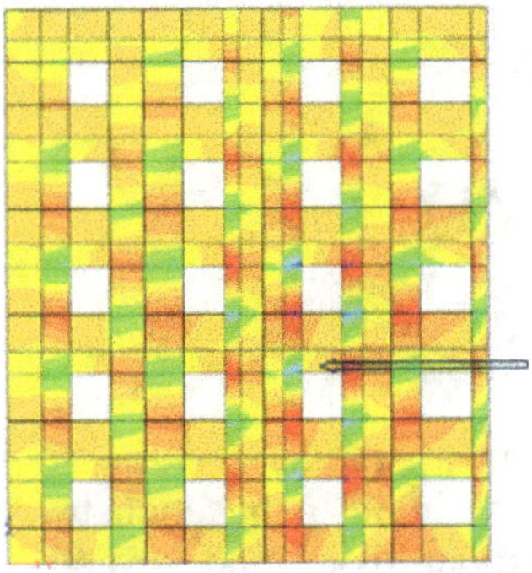

Fig. 9.19- Wall on axis **6-6**. With arrow the most stressed element W1053 is shown

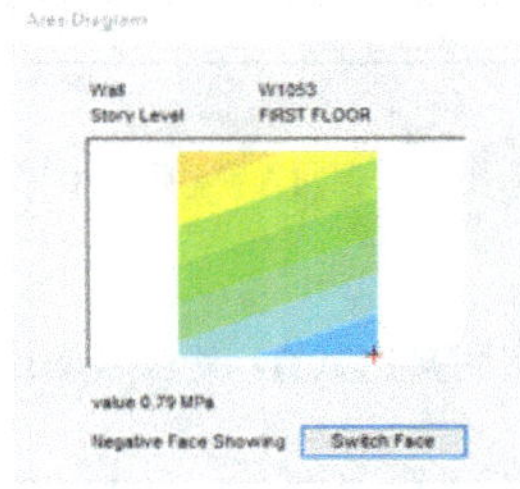

Fig. 9.20 Element W1053 detached from the axis 6-6

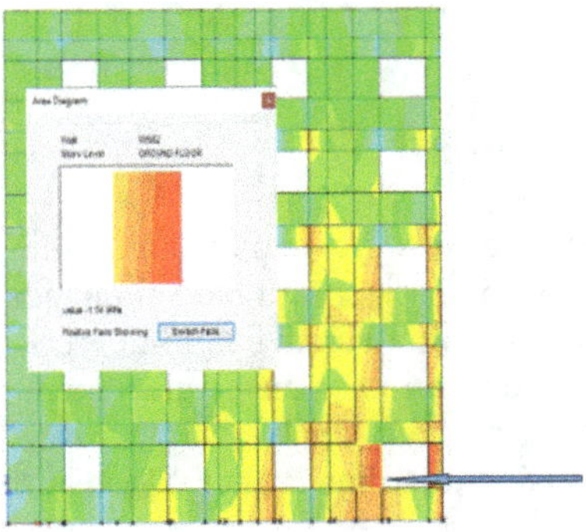

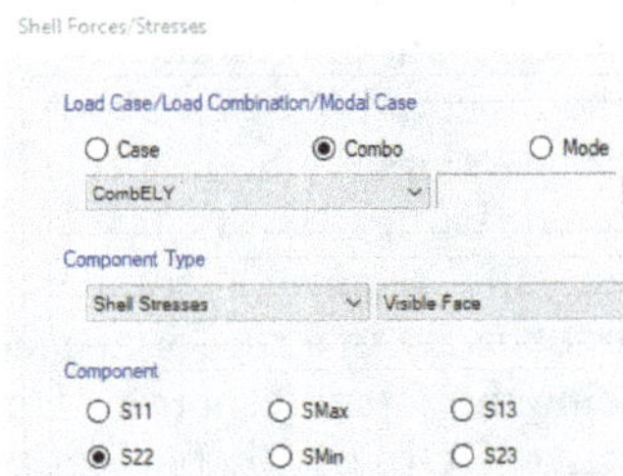

Fig. 9.21- The most stressed element W582

The maximum value of the stress in this axis is **S1-1** in tension, with a value of 0.79 MPa, while [S] in tension must be ≤ 0.2 MPa.
In the ELY combination, the wall on axis **6-6** appears to be at risk, the stress S_{22} =-1.54 N/mm^2, in compression, is much greater than the allowable stresses of the masonry.

Tab. 9.7- Table with recommendations for [σ] allowed in tension

Tension-type	**Mortar type**	
	M3, M4	M1,M2
Pulling perpendicular to the section of the masonry	2,0 kg/cm^2	2,5 kg/cm^2
Pulling parallel to the section of the masonry	3,9 kg/cm^2	5,0 kg/cm^2

CHAPTER 10 - NONLINEAR ANALYSIS

10.1 - METHOD of equivalent frames

As highlighted in previous chapters, a significant influence on the analysis of masonry, particularly in the nonlinear domain, was the initial publication by Tomasevic (Tomaževic, 1978, "**POR Method**"). Based on this method, several numerical programs have been developed that analyze models divided into macro-elements, such as **piers and spandrels**, under the method's recommendations and limitations. The adoption of performance-based earthquake engineering concepts has resulted in the implementation of various static nonlinear procedures for the seismic assessment of buildings.
- coefficient method (ASCE, 2000),
- capacity spectrum method (ATC, 1996; Freeman, 1998),
- N2 method (Fajfar, 1988),

They generally require a comparison between the seismic demand and the building's capacity in terms of displacements. This comparison can be conducted by replacing the actual response of the building with an equivalent single-degree-of-freedom (SDOF) oscillator (Shibata, 1976). The definition of the displacement capacity can be determined through a **nonlinear** static analysis (pushover analysis), which involves applying linearly increasing distributed lateral loads to the structure to simulate seismic forces. **[14]**

As for masonry STRUCTURES, this can be achieved in several ways:

□ - The **first** approach is represented by the finite element method, where the masonry construction elements (units and mortar) are discretized into a specified number of finite elements. Suitable nonlinear constitutive laws are applied, making the model a powerful analysis tool. However, its drawbacks include a high computational burden, which may render its use impractical, as well as potential grid dependence and the challenge of accurately calibrating the input parameters.

□- A second approach is based on the use of "equivalent frames," a model commonly employed by structural engineers. The structure is represented by a set of vertical and horizontal elements (piers and spandrels).

- The piers and spandrels are connected by **rigid joints**, and each element is modeled using appropriate constitutive laws to account for mechanical nonlinearity. This approach introduces significant simplifications, so its accuracy depends on the compatibility between the adopted assumptions and the actual structural problem. **[14]**

10.2- AMQUAKE (ATENA masonry earthquake)

Basic concepts related to the use of nonlinear building analysis 77/5

The program helps engineers design safe masonry buildings in seismic regions. This development was initiated by the introduction of Eurocodes 6 and 8 in Europe, starting in March 2010, which apply strict rules for masonry buildings in seismic regions. Am-Quake is a software program for the seismic assessment of masonry buildings, which uses **pushover analysis** and the **equivalent** frame method, based on macro-element discretization and its fundamental idea of the "floor mechanism" approach, to checking the seismic safety of masonry buildings, while maintaining concepts and idealizations that are familiar to the engineer.
In Am Quake, ***v6.0***, the floor slabs (slabs) are considered rigid, horizontal structures, transmitting the lateral (earthquake) load to the retaining walls, while the walls are automatically equivalent to frame-type elements, with non-linear behavior. This relationship is created to approximate the behavior and damage of wall panels as close to reality as possible.
The results of Am-Quake are extended over a wide range of validity and can be compared with those of a more sophisticated analysis. In addition, the program also supports the static analysis of masonry buildings according to **Eurocode 6**.

A typical frame model of walls is presented as follows:
Each wall panel is divided into macro elements named Piers and Lintels (Spandrels), modeled by beams and columns with non-linear behavior connected by rigid joints. At the joints of these elements, a space with plastic behavior is defined, with a distance accepted according to EC and their function is to transfer static and kinematic variables between the ends of the elements and the joints (Generation of Plastic Hinges).
As mentioned above, first the basic material parameters are set, then the spectrum data is checked and finally the geometric model is created. The program creates the 3D object, which is ready for analysis. Later the basic loads are applied. Also, the pushover load is performed according to the program's user manual.
The load combination for the push-over analysis is:

$$\gamma_{Gd}\, G + \psi_{EL}\, Q_L + \psi_{Es}\, Q_s \tag{10.1}$$

where: $\psi_{EL} = \varphi_L\, \psi_2,$

$\psi_{ES} = \varphi_S\, \psi_{2,s}$

Tab 10.1- Clarification of the terms used by the AmQuake program

Pushover Gamma Gp	Multiplier $\gamma G,p$ for the permanent load G for pushover analysis. In most national standards, this multiplier should be equal to 1.0.
Static Gamma Gs	Multiplier γGs - for permanent load G for static analysis
Pushover Phi_ L	Multiplier φ_L - for the temporary load QL for pushover analysis.
Static Gamma_QL	Multiplier $\psi_{2,s}$ - for snow load QS for pushover analysis
Pushover Psi_2L	Multiplier ψ_{2L} - for temporary load QL for pushover analysis
Pushover Phi_S	Multiplier φ_S - for snow load QS for pushover analysis
Pushover Psi_2S	Multiplier $\psi_{2,s}$ - për ngarkesën e borës Q_S për analizën pushover
Static Gamma Qs	Multiplier γ_{QS} - for snow load QS for static analysis
Gamma_M	Partial safety factor γ_M for masonry
Gamma_C	Partial safety factor γ_C for concrete
Gamma_S	Partial safety factor γ_S for steel

While the load combination for static analysis is:

$$\gamma_G G + \gamma_{QL} Q_L + \gamma_{QS} \psi_{0,s} Q_S \tag{10.2}$$

The most important step before starting the analysis is the automatic conversion or equivalence of the real building to the simplified static model, according to the principles expressed above. It should be noted that the partial safety factors for the materials specified above are used only in the static analysis. The pushover analysis is performed using the average material properties, i.e. the above factors are automatically set by the program to the value 1.0 for the pushover analysis

The most important step before starting the analysis is the automatic conversion or equivalence of the real building to the simplified static model, according to the principles expressed above. It should be noted that the partial safety factors for the materials specified above are used only in the static analysis. The pushover analysis is performed using the average material properties, i.e. the above factors are automatically set by the program to the value 1.0 for the this analysis.

10.3- EXAMPLE of schematizing an elementary structure for the Am Quake program

An important step in preparing the data for analysis with Am Quake is the idealization of the structure, which is based on the theory of equivalent frames. The process of connecting individual walls and windows is demonstrated using the figure below. Let us assume the modeling and interconnection of Wall1, Wall2, and Window. They are modeled by beam elements. The details of Wall 1,2 and Window 1 are shown in the next figure. The red lines in the figure below represent beam elements and the gray lines indicate connections between them. The retaining walls are modeled by vertical columns.

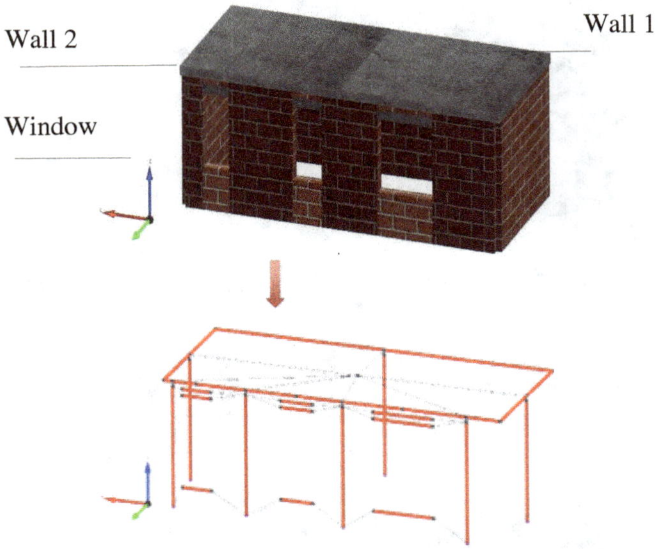

Each window wall is typically modeled with two beams, one for the parapet and one for the part of the wall above the lintel. The horizontal beams in the model are also used for the b/a rings in the basement floors.

Modeling on ATENA Masonry Quake – Version 6.0.0

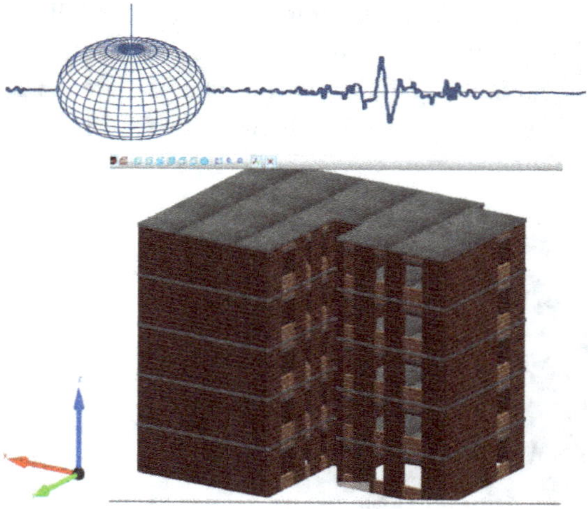

Fig. 10.2- 3D view of building 77/5 modeled for AM Quake

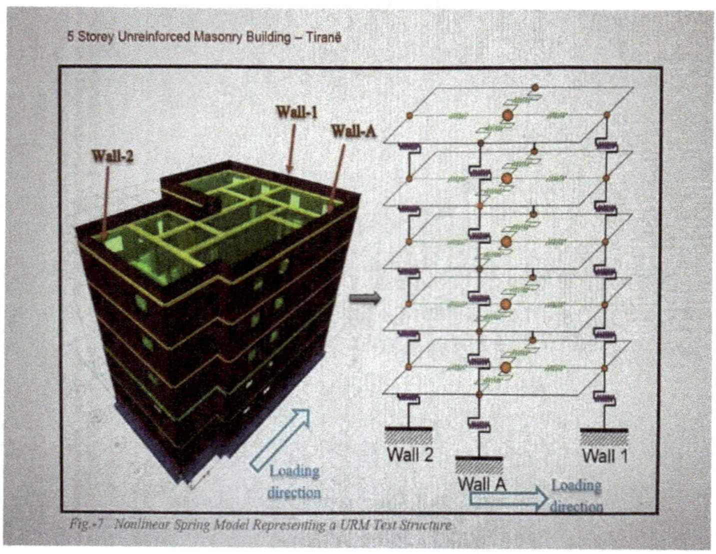

Fig. 10.3 – *Spring model representing the URM test structure*

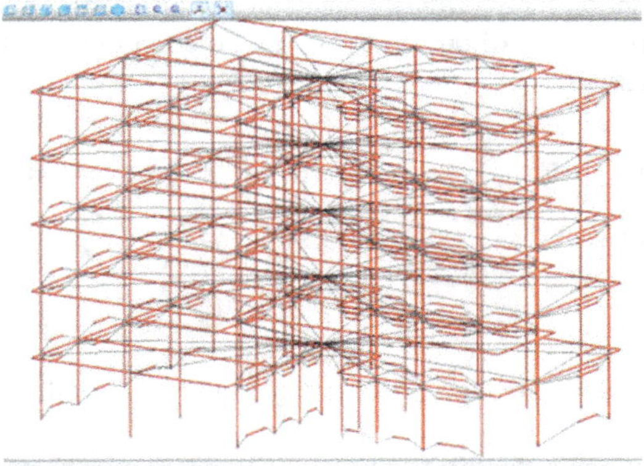

Fig. 10.4 - Structure in analysis modeled with frame elements. **The grid** (nodes, elements and connections)

ANALYSIS PARAMETERS FOR SOME COUNTRIES

a- Cyprus

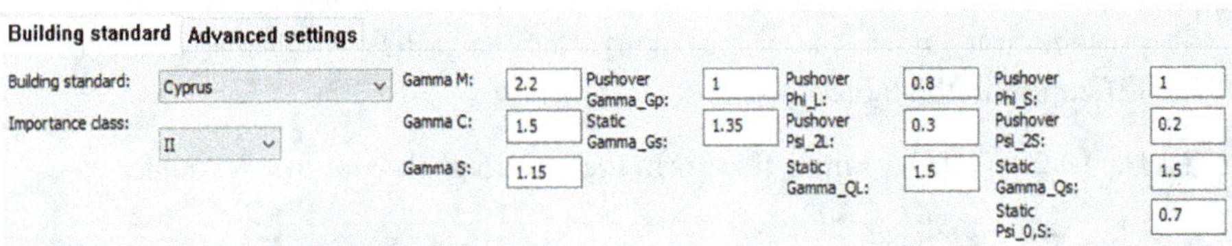

b- Romania (Eurocode)

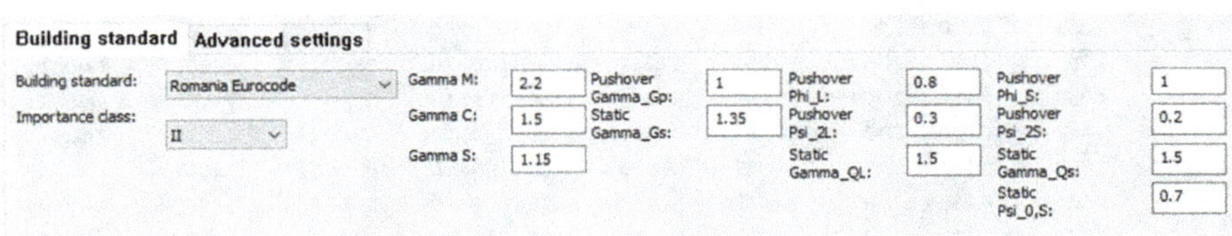

c- Bulgaria

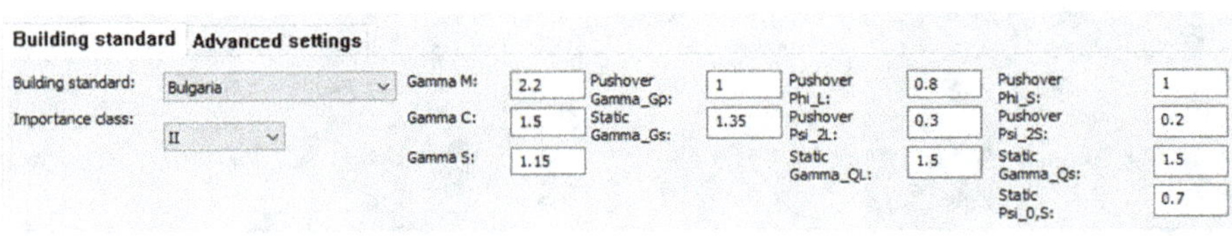

Fig. 10.5 - Comparison of analysis parameters for several countries

As seen above, these parameters are the same for all countries that the program has included. For brevity of presentation, not all countries that the program references are shown.

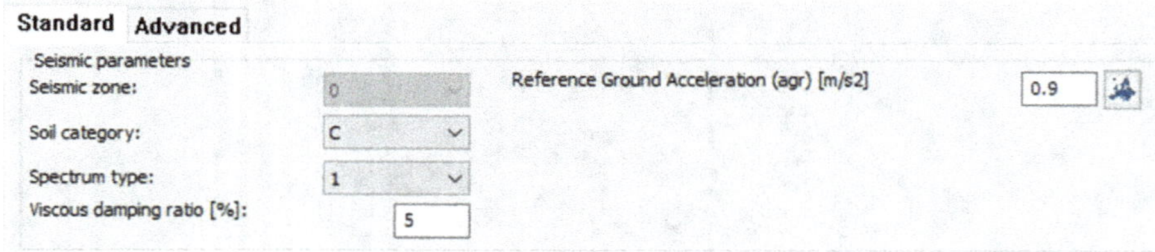

Fig. 10.6 - Soil category, spectrum type and soil acceleration.

Since this table is again the same for all countries listed in the program, for brevity of reference they have not been cited. Our country is not on the list of countries that the program has included, therefore, for the respective analysis, Cyprus has been chosen.

The following table gives the dimensions of the column-walls (Piers), the coordinates of the centers of gravity of each wall in particular, the angle they form with the main axes. Only a part of these column-walls is given, for brevity of presentation, since there are many and each has the same logic of its identification in the table.

TAB. 10.2: Walls/Columns that form the building's layout in AM quake

All #	Length [m]	Width [m]	Height [m]	Xg	Yg	Z0	Rotation	In plane eccentricity [m]	Out of plane eccentricity [m]	Material
1	8.850	0.250	2.700	14.489	14.251	0.000	0.000	0.000	0.000	muri 25
2	1.350	0.250	2.700	10.064	13.576	0.000	90.000	0.000	0.000	muri 25
3	1.260	0.250	2.700	10.064	10.851	0.000	90.000	0.000	0.000	muri 25
4	1.240	0.250	2.700	10.064	9.021	0.000	90.000	0.000	0.000	muri 25
5	0.850	0.250	2.700	10.064	7.398	0.000	90.000	0.000	0.000	muri 25
6	8.850	0.250	2.700	14.489	10.596	0.000	0.000	0.000	0.000	muri 25
7	3.280	0.250	2.700	15.019	5.336	0.000	90.000	0.000	0.000	muri 25
8	11.250	0.250	2.700	13.291	6.976	0.000	0.000	0.000	0.000	muri 25
9	5.380	0.250	2.700	16.226	0.246	0.000	0.000	0.000	0.000	muri 25
10	2.660	0.250	2.700	15.019	1.573	0.000	90.000	0.000	0.000	muri 25
11	1.340	0.250	2.700	12.288	0.246	0.000	0.000	0.000	0.000	muri 25
12	1.340	0.250	2.700	10.248	0.246	0.000	0.000	0.000	0.000	muri 25
13	1.400	0.250	2.700	7.458	0.246	0.000	0.000	0.000	0.000	muri 25
14	1.670	0.250	2.700	4.802	0.246	0.000	0.000	0.000	0.000	muri 25
15	4.090	0.250	2.700	3.966	2.293	0.000	90.000	0.000	0.000	muri 25

While the following table gives the dimensions of the Doors-Windows, the coordinates of the centers of gravity of each space in particular, the angle they form with the main axes. Only a part of these walls-columns is given, for brevity of presentation.

TAB. 10.3: Doors/Windows that complement the building scheme in AM quake

All #	Length [m]	Width [m]	Height [m]	Xg	Yg	Z0	Rotation	Parapet height	Window/door height	Parapet material	Lintel material
31	1.300	0.250	2.700	2.484	10.096	0.000	0.000	0.750	1.500	muri 25	Lintels 14
32	0.700	0.250	2.700	4.791	10.096	0.000	0.000	0.750	1.500	muri 25	Lintels 14
33	0.630	0.250	2.700	6.952	10.096	0.000	0.000	0.750	1.500	muri 25	Lintels 14
34	1.420	0.250	2.700	0.366	3.631	0.000	90.000	0.750	1.500	muri 25	Lintels 14
35	1.120	0.250	2.700	6.198	0.246	0.000	0.000	0.750	1.500	muri 25	Lintels 14
36	1.420	0.250	2.700	8.868	0.246	0.000	0.000	0.750	1.500	muri 25	Lintels 14
37	0.700	0.250	2.700	11.268	0.246	0.000	0.000	0.750	1.500	muri 25	Lintels 14
38	0.580	0.250	2.700	13.248	0.246	0.000	0.000	0.750	1.500	muri 25	Lintels 14
39	1.120	0.250	2.700	18.915	2.481	0.000	90.000	0.750	1.500	muri 25	Lintels 14
40	1.120	0.250	2.700	18.915	5.901	0.000	90.000	0.750	1.500	muri 25	Lintels 14
41	1.120	0.250	2.700	18.915	9.221	0.000	90.000	0.750	1.500	muri 25	Lintels 14
42	0.920	0.250	2.700	18.915	12.517	0.000	90.000	0.750	1.500	muri 25	Lintels 14
43	1.420	0.250	2.700	10.064	12.191	0.000	90.000	0.750	1.500	muri 25	Lintels 14
44	0.580	0.250	2.700	10.064	9.931	0.000	90.000	0.750	1.500	muri 25	Lintels 14
45	0.580	0.250	2.700	10.064	8.111	0.000	90.000	0.750	1.500	muri 25	Lintels 14
77	1.300	0.250	2.700	2.484	10.096	2.900	0.000	0.750	1.500	muri 25	Lintels 14

Below are the data for the floors, according to the format required by the program, as well as the respective floor loads.

TAB. 10.4: Geometric data for the building floors

All #	Name	Area [m2]	Height [m]	Number of geometry	Number of loading areas	Z	RC Ring
46	Floor 0.000 - 2.900	217.380	0.200	12	1	2.700	RC Ring 15
92	Floor 2.900 - 5.800	217.380	0.200	12	1	5.600	RC Ring 15
138	Floor 5.800 - 8.700	217.380	0.200	12	1	8.500	RC Ring 15
184	Floor 8.700 - 11.600	217.380	0.200	12	1	11.400	RC Ring 15
230	Floor 11.600 - 14.500	217.380	0.200	12	1	14.300	RC Ring 15

TAB. 10.5: Loading data for the building floors

All #	Floor	Dead load [kN/m2]	Permanent load [kN/m2]	Pushover Gamma_Gp	Static Gamma_Gs	Life load [kN/m2]	Pushover Phi_L	Pushover Psi_2L	Static Gamma_QL	Snow load [kN/m2]	Pushover Phi_S	Pushover Psi_2S	Static Gamma_Qs	Static Psi_0,S
46	Floor 0.000 - 2.900	3.000	1.500	1.000	1.350	2.000	0.800	0.300	1.500	0.000	1.000	0.200	1.500	0.700
92	Floor 2.900 - 5.800	3.000	1.500	1.000	1.350	2.000	0.800	0.300	1.500	0.000	1.000	0.200	1.500	0.700
138	Floor 5.800 - 8.700	2.500	1.500	1.000	1.350	2.000	0.800	0.300	1.500	0.000	1.000	0.200	1.500	0.700
184	Floor 8.700 - 11.600	2.500	1.500	1.000	1.350	2.000	0.800	0.300	1.500	0.000	1.000	0.200	1.500	0.700
230	Floor 11.600 - 14.500	2.500	1.500	1.000	1.350	2.000	0.800	0.300	1.500	0.000	1.000	0.200	1.500	0.700

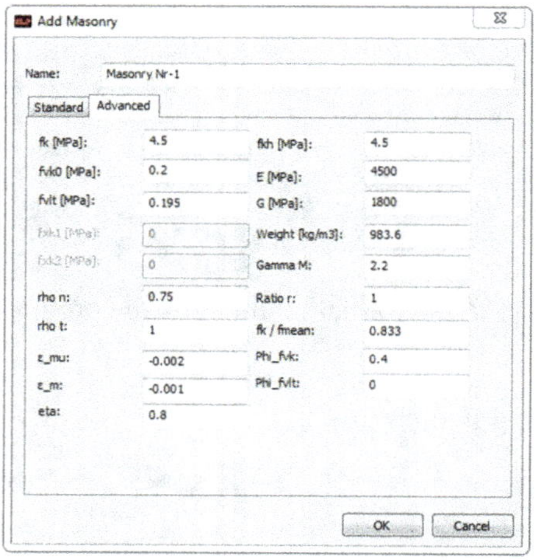

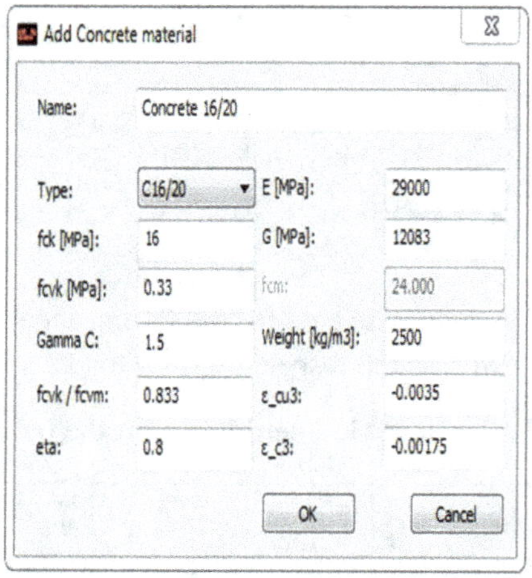

Fig.- 10.7. Data on bricks, Input in Am quake

Fig.- 10.8. Concrete data

10.4 - PUSHOVER ANALYSIS REPORT (Short Protocol)

Tab. 10.6 Step 2: push X and push Y (results)

Steps	Analysis	Limit states	DLS target [mm]	DLS capacity [mm]	ULS target x 1.50 [mm]	ULS capacity [mm]	Safety margin DLS [%]	Safety margin ULS [%]	Overstrength fact.	Max error [%]
	Direction X									
1	X+, exc. pos, uni	Pass	3.93	13.83	13.71	13.83	71.58	0.89	1.23	0.56
2	X+, exc. pos, tri	Pass	5.21	17.93	17.86	17.93	70.95	0.66	1.50	0.40
3	X+, exc. neg, uni	Pass	5.40	18.33	18.06	18.33	70.55	1.43	1.11	0.43
4	X+, exc.neg, tri	Pass	4.01	14.79	14.45	14.79	72.86	2.26	1.18	0.36
5	X-, exc. pos, uni	Pass	3.43	12.95	12.41	12.95	73.47	4.17	1.03	0.48
6	X-, exc. pos, tri	Pass	3.95	16.74	14.27	16.73	76.42	14.70	1.05	0.47
7	X-, exc. neg, uni	Pass	3.41	15.90	12.38	15.90	78.53	22.15	1.00	0.52
8	X-, exc.neg, tri	Pass	4.11	15.44	14.81	15.44	73.36	4.06	1.02	0.44
	Direction Y									
1	Y+, exc. pos, uni	Pass	22.34	66.00	67.04	67.10	66.15	0.089	1.35	0.25
2	Y+, exc. pos, tri	Fail	25.65	63.75	76.96	74.35	59.76	-3.50	1.59	1.57
3	Y+, exc. neg, uni	Fail	25.40	65.86	76.21	67.86	61.43	-12.31	1.41	0.31
4	Y+, exc.neg, tri	Fail	28.24	62.24	84.71	84.42	54.63	-0.35	1.60	2.94
5	Y-, exc. pos, uni	Pass	21.17	63.76	63.50	63.76	66.80	0.42	1.36	0.29
6	Y-, exc. pos, tri	Pass	23.57	61.83	70.71	71.42	61.88	1.00	1.42	0.34
7	Y-, exc. neg, uni	Pass	20.80	63.65	62.41	63.65	67.32	1.95	1.35	0.33

Tab. 10.7 Step 3 push X and push Y (results)

Nr	Analysis	Limit states	Max error [%]	Storey	RShift	Dls Criterion	Dls Step Id	Period T Norm.	Max ref. ground accel. [m/s2]	Ductility	Elastic displac[mm] = d_y
	Direction X										
1	X+, exc. pos, uni	Pass	0.712	1	0.002	3.519	10	0.336	0.905	2.355	6.232
2	X+, exc. pos, tri	Pass	5.683	1	0.0016	3.442	35	0.357	0.904	2.70	8.273
3	X+, exc. neg, uni	Pass	5.398	1	0.0021	3.395	10	0.394	0.910	2.313	8.412
4	X+, exc.neg, tri	Pass	3.841	2	0.0013	3.684	30	0.314	0.913	2.937	6.266
5	X-, exc. pos, uni	Pass	0.717	1	0.0016	3.769	8	0.314	0.925	2.579	5.329
6	X-, exc. pos, tri	Pass	0.613	1	0.0015	4.241	9	0.311	0.997	3.390	6.142
7	X-, exc. neg, uni	Pass	0.679	1	0.0021	4.657	7	0.313	1.062	3.201	5.275
	Direction Y										
1	Y+, exc. pos, uni	Pass	23.00	4	0.005	2.954	184	1.071	0.901	3.496	20.369
2	Y+, exc. pos, tri	Fail	37.28	5	0.005	2.485	228	1.049	0.870	3.877	23.864
3	Y+, exc. neg, uni	Fail	17.70	4	0.005	2.592	173	1.217	0.801	2.590	27.814
4	Y+, exc.neg, tri	Fail	19.94	4	0.0049	2.204	118	1.154	0.897	3.567	29.443
5	Y-, exc. pos, uni	Pass	0.858	4	0.0048	3.013	39	1.014	0.904	3.590	18.852
6	Y-, exc. pos, tri	Pass	8.401	5	0.0049	2.623	33	0.963	0.909	4.895	18.152

Tab. 10.8 - *Explanation of basic terms used in the program* **AM quake:**

DLS target	(Damage limite state), damage limitation condition, the displacement of the upper floor, in which inter-story movements, for each floor, are controlled, if they do not violate the damage limitation requirements. The inter-story drift for each floor must be less than the damage limit value;
DLS capacity	Displacement of the upper floor, when the mezzanine drift, for a given floor, exceeds the damage limit value. The mezzanine drift for each floor must be less than the damage limit value;
ULS target	(Ultimate Limit State) - the top floor displacements that can be expected for the structure and the given seismic demand. The structure must be able to withstand such top floor displacements without collapsing.
Safety margin DLS	Safety margin for DLS in % of DLS capacity
Safety margin ULS	Safety margin for ULS in % of ULS capacity
DLS Criterion	DLS Pushover Criterion, DLS Capacity/DLS Target
ULS Criterion	ULS Pushover Criteria, ULS Capacity/DLS Target
DLS Step Id	Step Id, when the inter-story damage limit is reached. The floor Id is shown in the Storey parameter
Period T Norm	The period of oscillation of an idealized structural system with one degree of freedom
Max reference ground	The maximum possible reference ground acceleration for the analyzed acceleration model. For the given structural geometry, loading direction, and selected spectrum type, it provides the value of the maximum reference ground acceleration that the structure will be able to survive.
Ductility	This value is used to determine the q factor for the analyzed structure. It is the ratio of the maximum shear force value to the shear force value at the base, when the first plastic hinge develops in one of the vertical elements.

Seismic parameters

Seismic zone:
Soil category: C
Spectrum type: 1
Damping ratio: 5.000
Ultimate acceleration: 2.70
Damage acceleration: 0.27
Soil factor (S): 1.150
Period B: 0.200

Period C: 0.60
Period D: 2.000
Dynamic amplif. factor beta0: 2.500

** **CAPACITY CURVES (for some of the analysis steps)**

PUSH- ACCORDING TO DIRECTION X

Step-1 * Push Analisis -X/1 (Limit state -> **Pass**!)

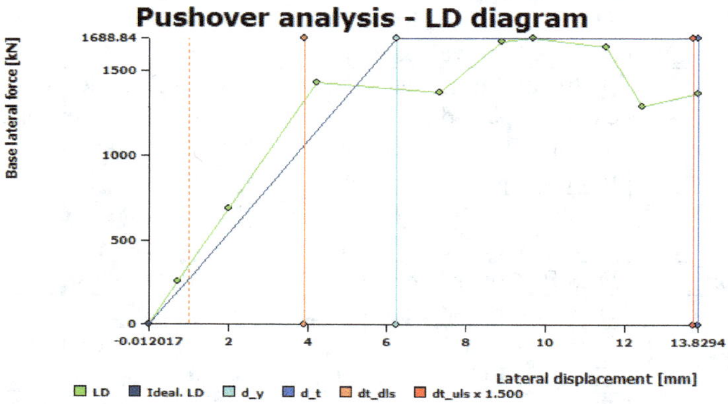

Fig.-10.9 - Push-X Step-1

Ld diagram	- lateral displacement diagram
d_ y	- Limit displacement
d_ t	- targeted displacement
dt_ dls	- targeted displacement for DLS
dt_ uls	- targeted displacement for ULS (ultimate limite state)

Step-2 * Push Analisis -X/2 (Limit state -> **Pass**!)

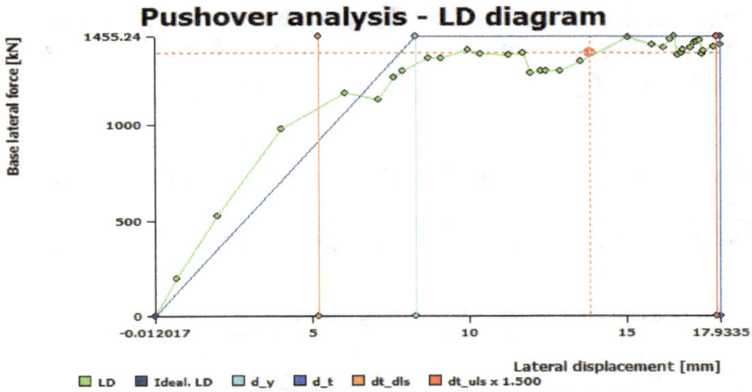

Fig.- 10.10 - Push-X Step-2

Step-3 * Push Analisis -X/3 (Limit state -> **Pass**!)

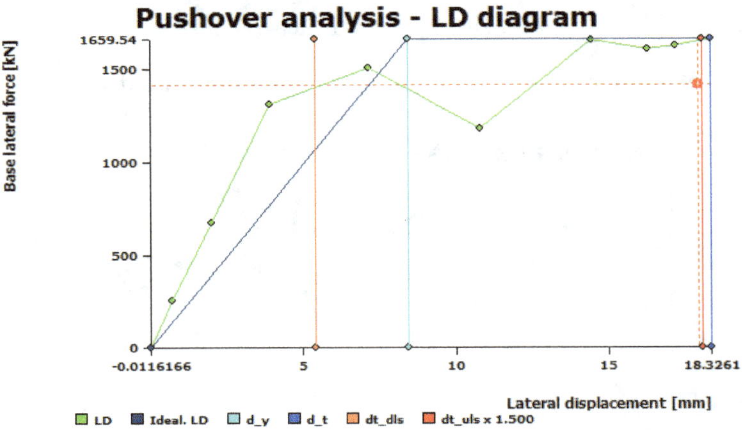

Fig. 10.11 - Push-X Step-3

Since all the PUSH x loading steps are subject to the respective load, we are not presenting the other pushover diagrams, which are more or less repeated. However, we are presenting below the PUSH y diagrams, which "collapse" the building. Thus, the building does not have the proper capacity, and it collapses in steps 2, 3, and 4 of this reaction.

PUSH- ACCORDING TO THE Y DIRECTION

Step-2 * Push-Y/2 Analysis (Limit State -> Collapse!)

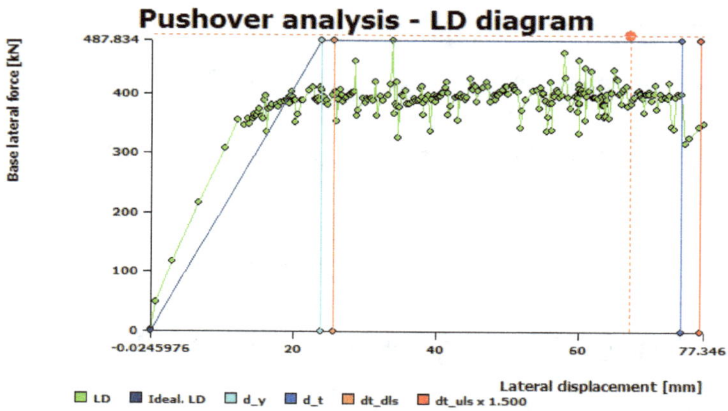

Fig. 10.12 - Push-Y Step-2

Step-3 * Push-Y/3 Analysis (Limit State -> Collapse!)

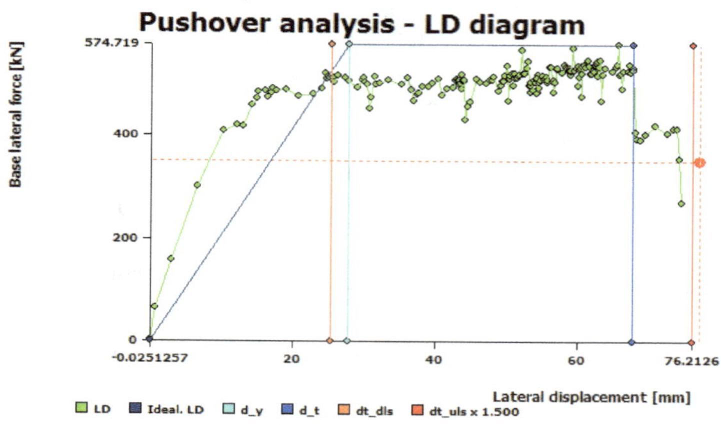

Fig. 10.13 - Push-Y Step-3

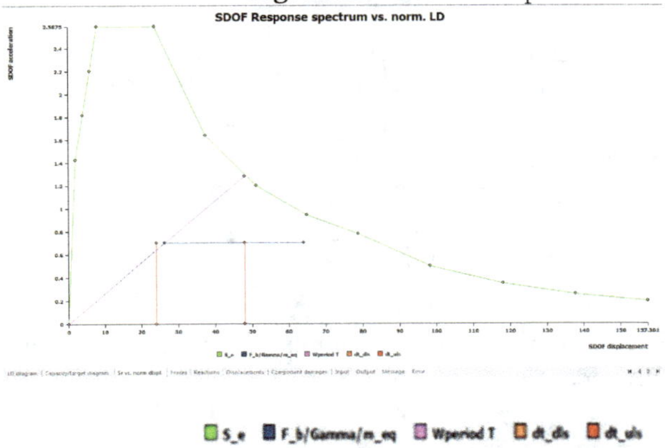

Fig. 10.14 - Push-Y, Step -2.
Sa-Sd pushover curve. (Response spectrum alongside with displacement spectrum)

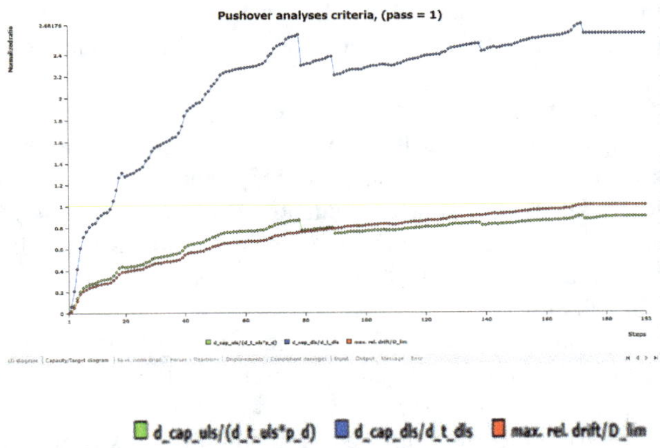

Fig. 10.15 - Push-Y, Step -2. Capacity diagrams and permitted limits according to the EC

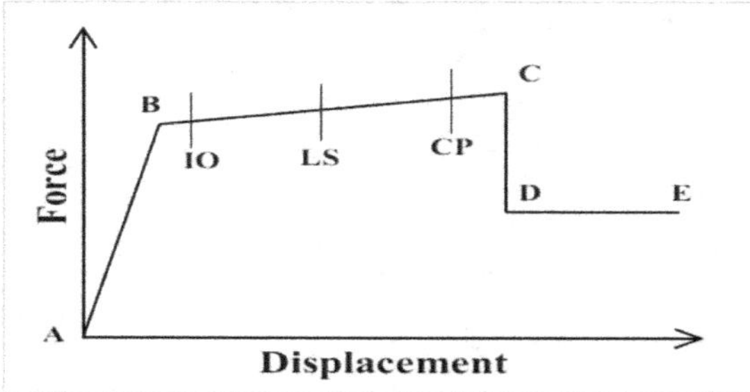

Fig. 10.16 - Force-displacement curves and limits for pushover analysis according to the Eurocode.

10.7 - RESULTS ANALYSIS:

The summary table presents the results from all stages of the Pushover analysis, focusing specifically on the most critical aspects of the building's gradual loading and response. The gradual loading using Push-X successfully navigates all steps, whereas the loading using Push-Y does not, indicating that the building experiences global collapse in this direction. This happens despite some individual elements in the local plan demonstrating additional capacity. By analyzing the Push-Y capacity curve, step 3, we can make a detailed comment on the corresponding capacity curves versus seismic demand: The Push-Y capacity curve at step 3 exhibits significant fluctuations, although it does not indicate collapse at this stage. It is important to highlight that the building under analysis has an unreinforced masonry (**URM**) typology, where each joint or element responds in a discrete, plastic manner. This behavior contrasts with reinforced concrete (RC) structures, where elements and joints typically exhibit similar elastic or elastoplastic behavior throughout all phases, resulting in smaller fluctuations in the capacity curve.

Analyzing the curve reveals that each node exhibits varying forms of post-elastic hardening and softening. This variability can present challenges when utilizing synthetic inelastic spectra. With repeated Push-Y loading, the curves tend to converge. In this scenario, the ductility requirement aligns with the capacity. Therefore, in theory, the capacity remains contingent upon the shape of the nonlinear curve.

Consequently, in theory, the capacity remains dependent on the shape of the nonlinear curve. At the target displacement limits determined by the A.M. Quake program (based on EC-8), different parts of the structure in the local plane do not withstand the impacts caused by the Push-Y loading, resulting in irreversible plastic collapse. The internal potential energy fails to meet the spectral demand, leading to a stage where the structure undergoes global collapse, as assessed according to limit states design.

Regarding the Pushover nonlinear static analysis, we have concluded that:

- *The structure experiences issues and collapses after only a few steps in the Y-direction, which is the most problematic aspect of the structure, while all gradual loading steps in the X-direction are successfully completed.*
- *In the steps according to the Push-Y loading, we have noticed that the structure has a low capacity in the face of the seismic demand (design spectrum) and a high displacement;*
- *The safety margin for the "Live Safety" level, referred to these 4 steps, where the building collapses, is very low (related to the ULS, the ultimate limit state).*

CHAPTER 11

ATTEMPTS AND RECOMMENDATIONS for improving the BUILDING parameters

To enhance the two main parameters of the building, namely the period and the locations, several attempts were made to find solutions for modifying these parameters.

The attempts undertaken are outlined below.

11.1 - ATTEMPT 1 [LAYERED MASONRY]

Based on the layered masonry option offered by Etabs, it was assumed that one side of the masonry is reinforced with a 75 mm thick concrete lining. A variant featuring four integration points was selected for analysis.

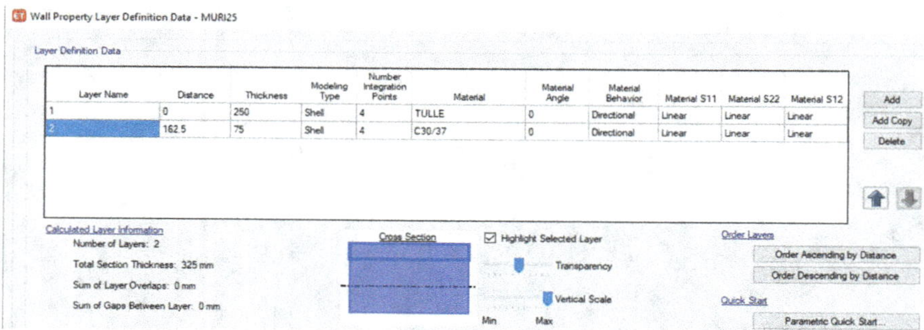

Fig. 11.1 - Brick masonry coating with C30/37 concrete, 75 mm thick.

The results of this attempt are given below:
The period in mode 1 is (0.709 sec), while the displacements are: Uy= 104.27mm and Ux = 22.18 mm!!

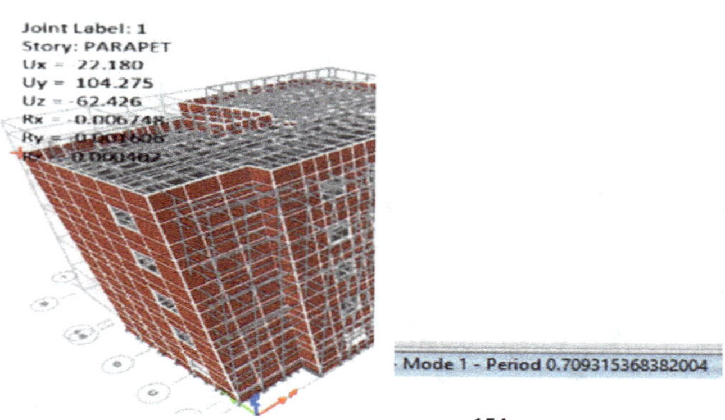

Fig. 11.2- Periods and displacements at the top of the building in attempt 1

The improvement is insignificant. The building continues to be flexible, the displacements in the plastic phase are 10.42cm * 1.5(ductility) = 15.7 cm, while they should have been 7 cm!!!

11.2- ATTEMPT 2

In this attempt, it was assumed that the masonry would be covered with 75 mm thick concrete on both sides.

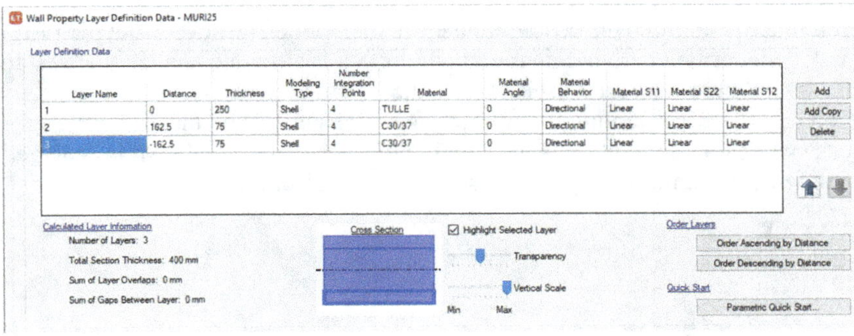

Fig. 11.3 - Concrete coating C30/37, thick 75 mm on both sides of the masonry

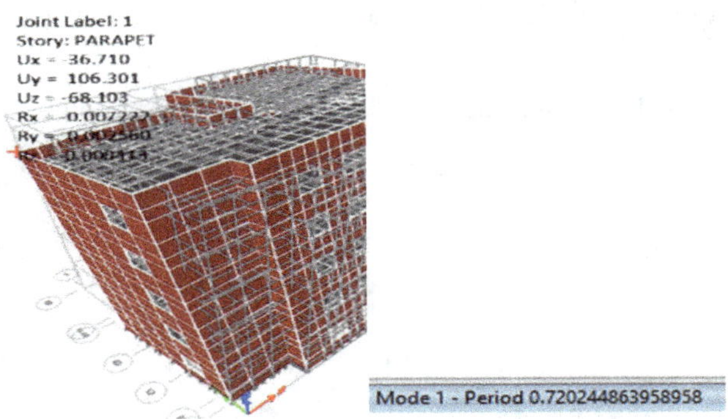

Fig. 11.4 - Periods and displacements at the top of the building in attempt 2

So, the period T is 0.720 sec, the displacements are: Uy =106.30 mm and Ux - 36.71 mm. Our analysis indicates that adding two concrete layers of C30/37 to the walls, negatively impacts performance, compared to the scenario with a single layer. Although, the increases in both, period and displacements are minimal, they are still notable. Additionally, we emphasize that altering the integration point parameter does not affect either the period or displacements of the building. In a scenario where only the first two floors were layered, there was again no significant change; the displacements remained at 103 mm, while the period adjusted to be

Mode 1 - Period 0.651155152338637

Per reference po citojme me poshte edhe nje here perioden dhe cvedosjet per rastin kur nuk behet asnje nderhyrje ne murature, dmth kur ajo eshte analizuar sic eshte ndertuar, me mure tulle me trashesi t= 38 dhe 25 cm (situata reale).
For reference, we are quoting below once again the period and the differences for the case when no intervention is made in the masonry, that is, when it is analyzed as built, with brick masonry, thicknesses t= 38 and 25 cm (real situation).

Mode 1 - Period 0.688960526380029

Fig. 11.5 - Displacements and periods for the original building

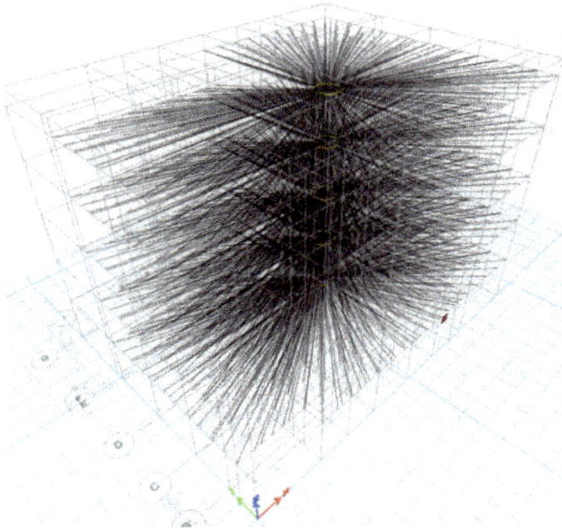

Fig. 11.6 - 3D visualization of the spatial relationships between wall elements and the building's center of mass, as generated by the ETABS program.

11.3- ATTEMPT 3 [Buiding with concrete walls C30/37]

a- NORMAL CONCRETE

a1- For a wall thickness of 10 cm, we have:
Period T = 0.537 sec, displacements Uy = 67.25 mm:

Mode 1 - Period 0.536860966498619

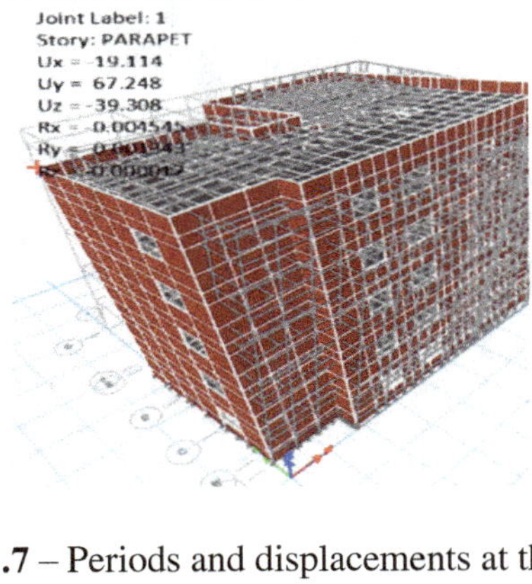

Fig. 11.7 – Periods and displacements at the top of the building for walls with t= 10cm

a2- For a wall thickness of 20 cm, we have the period:

Mode 1 - Period 0.613190237527508

Period T = 0.613 sec, displacements Uy = 86.12 mm: For the case with a wall arameters show a significant increase compared to the hypothetical 10 cm walls! The periods increase by 0.613/0.538 = 1.14 times, while the displacements increase by 86/67 = 1.28 times, which approximately corresponds to the square of the ratios of the periods!

b- <u>LIGHTWEIGHT concrete walls (hypothetical case)</u>

It is assumed that the only characteristic of concrete that changes is its self-weight. Therefore, a building with walls made of C 30/37 concrete, with thicknesses of 10 cm and 20 cm, but with a weight of 10 kN/m³, has been analyzed. As noted above, this is an idealized case intended solely for study purposes.
The results of this analysis are:
b1- For the case of concrete walls with t = 10 cm, we have:

Period T = 0.481 sec, displacements Uy = 53.04 mm

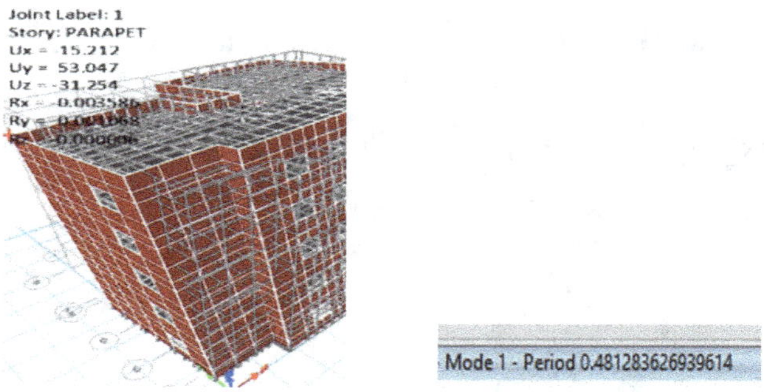

Fig. 11.8 – Periods and displacements at the top of the building for walls with t= 10cm (lightweight concrete)

b2- For the case of concrete walls with t = 20 cm, we have:
Period T = 0.514 sec, displacements Uy = 61.10 mm

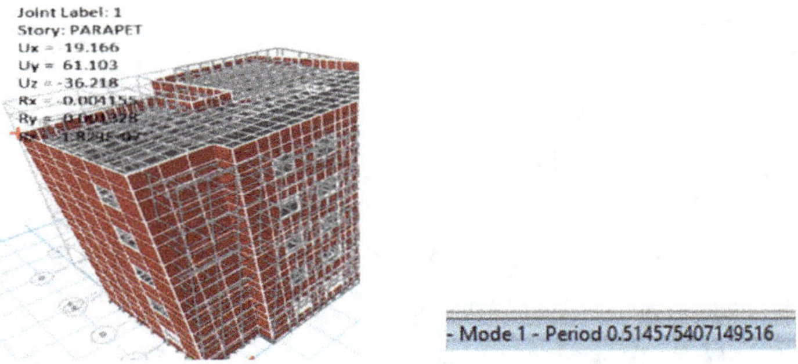

Fig. 11.9 – Periods and displacements at the top of the building for walls with t= 20cm (lightweight concrete)

The ratio between the periods is: 514/481 = 1.07, while the ratio of the displacements is 1.15!

11.4- ATTEMPT 4

[BUILDING WITH **CONCRETE WALLS** C30/37, + concrete columns on the first 2 floors]

Below is the reaction of the building, which features walls made of C30/37 concrete with a thickness of 20 cm, along with additional columns made of 30x30 concrete reinforced with 8 bars of diameter 16 on the first two floors. It is noteworthy that when 11 columns are placed on these two floors, the building's parameters remain unchanged; however, when 5 more columns are added, the target values in the periods are immediately achieved, while the displacements are limited to a maximum, resulting in the building becoming almost rigid
This analysis aimed to compare, discuss, and demonstrate that the results of the brick masonry version are reliable, as the same outcomes are achieved when the building is constructed with masonry and when columns are incorporated. I emphasize that the analysis with brick walls was conducted first, despite being presented at the end due to the study's objectives. This focus was chosen because the results for concrete masonry are regarded as more reliable, owing to the superior physical properties of concrete.

The results of this analysis are given below:

A- VERSION WITH 11 COLUMNS (only on 2 floors)

The period, in the first mode, in this case will be:

Mode 1 - Period 0.613621063681679

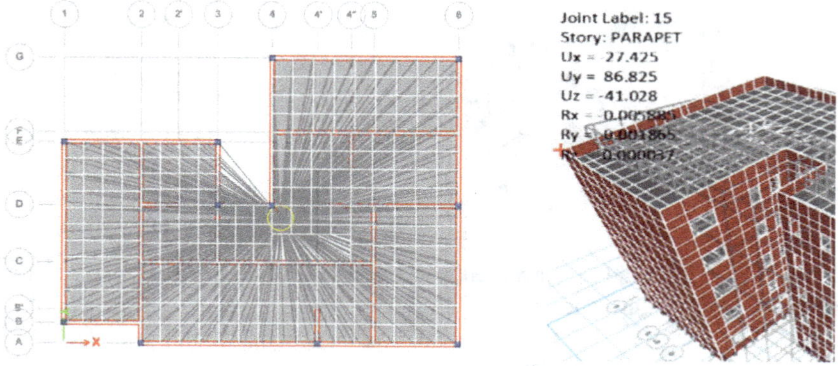

Fig. 11.10 – Typical floor plan with 11 columns and the locations for the ELY combination

B- 16 COLUMNS VERSION (only on 2 floors)

The period in this case is:

Mode 1 - Period 0.435000578491384

As can be observed, altering the number of columns results in a significant change in the parameters.

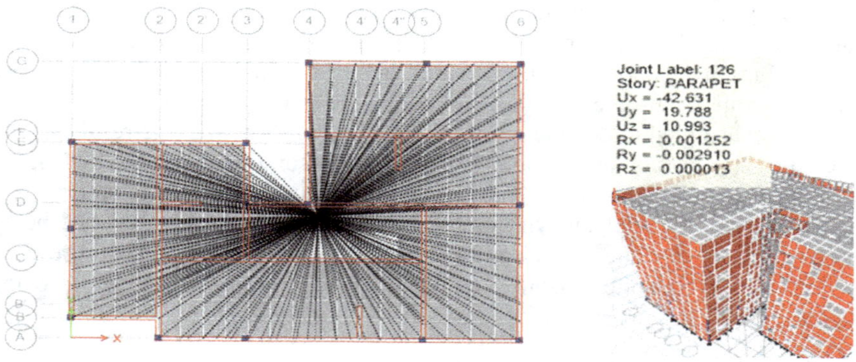

Fig. 11.11 – The **16** columns typical floor plan and the displacements at the top of the building

11.5- ATTEMPT 5

1- It was assumed that the masonry possesses a high modulus of elasticity,estimated to be twice the value recommended by Eurocode 6, resulting in an assumed modulus of elasticity of E = 2200 MPa. However, this remains a hypothetical scenario, as enhancing the properties of the existing masonry is virtually impossible; no feasible methods exist to improve the quality of the mortar and brick, the two primary components of the structure.

a- The impact of this change is insignificant in reducing the period; it only drops to the level T = 0.566 sec.

Mode 1 - Period 0.566201858962981

b- However, the displacements were improved for the ELY combination, the resulting displacement values, for the most unfavorable scenario, are as follows

c- The stresses along the 2-2 axis, identified as problematic, for the most unfavorable ELY combination are as follows:

Max = 78.257 at [7.17, 0, 14.6]; Min = -59.867 at [9.7, 13.87, 13.2222]

i.e. 3 times less than the values of these parameters for the building in its

Max = 0.21 at [3.6, 6.5756, 13.8]; Min = -0.59 at [3.6, 7.34, 0]

ATTEMPT 6

Some authors advocate for using fiberglass to reinforce masonry buildings, citing its lightweight properties, high tensile strength, and resistance to corrosion as significant advantages that enhance structural integrity and longevity.

a- This attempt did not yield any results, as evidenced by the significant and inestimable changes in periods and displacements, illustrated in Figure 11.13..

b- The stresses S2-2, in axis 6 are as follows :

Max = 5.75 at [18.42, 8.495, 7.7778]; Min = -10.9 at [18.42, 13.4188, 2.1778]

Thus, we have an increase in stresses, which was expected due to the stress-absorbing capacity of the material of Fiberglass.
In my opinion, reinforcing with FRP is not only associated with minimal parameter changes but is also practically unfeasible due to the high costs involved, particularly for large-scale structures such as 5-6-story residential complexes, we are analyzing.

THE MAIN CHARACTERISTICS OF FIBERGLASS [18]

PROPERTIES	ASTM Standart	22* C
Elastic constants		**GPa**
Longitudinal module E_L	D3039	53-59
Transverse module E_T	D3039	16-20
Axial shear modulus G_{LT}	D3518	6 -9
Poisson's Ratio **m** $_{LT}$	D3039	0.26-0.28
Resistance		**MPa**
Longitudinal traction $F^T{}_L$	D3039	1590-2000
Longitudinal pressure $F^C{}_C$	D3410	690-1240
Transverse traction $F^T{}_T$	D3039	41-82
Transverse pressure $F^{CH}{}_T$	D3410	110-200

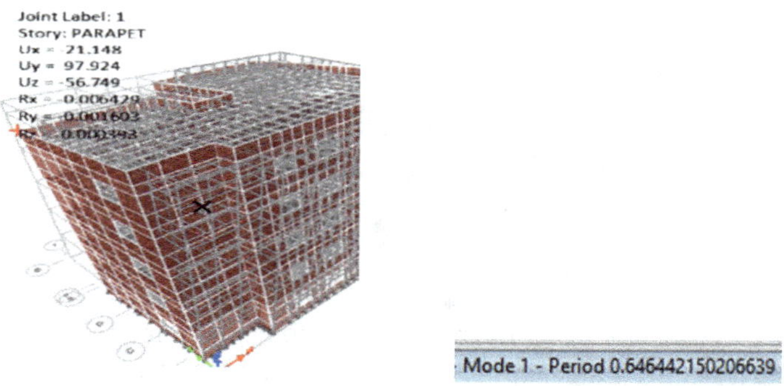

Fig. 11.12 – Periods and locations at the top of the building

Additionally, the rough surfaces of masonry, influenced by deep mortar joint gaps, necessitate significant material consumption for proper application and create breaking points that disrupt uniform force transmission between the masonry and the FRPs, further complicating the implementation process.

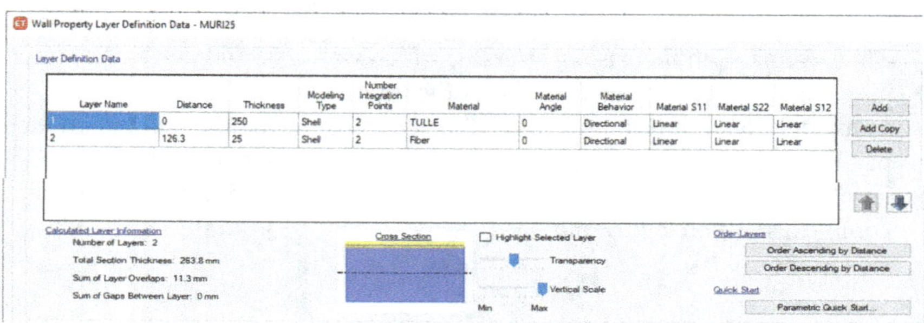

Fig. 11.13 – The fiberglass layer on the outer masonry wall

11.7- ATTEMPT 7

Given the results achieved, I decided to conclude my efforts, designating this as the final attempt, which aligns with the primary objective of this doctoral thesis: to identify a technically sound and economically viable method for upgrading existing masonry buildings to be resilient against potential seismic events. This attempt involved installing several columns on the first two floors of the masonry structure. Their positioning is shown in the figures below, in plan and 3D. With this attempt, in the Etabs model, the following parameters were changed: a- The period in the first modal form was reduced to the value T= 0,352 sek.
Let's remember that the allowed period for the building is [T] = 0.365 sec

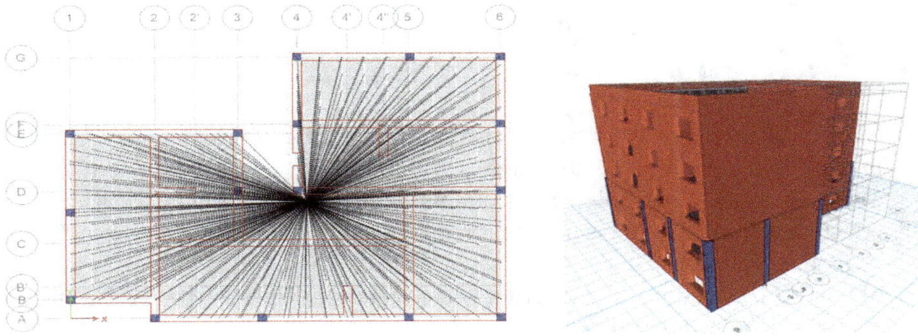

Fig. 11.14 - Column position in plan and in 3D

a- The maximum displacements for the ELY combination were reduced to the value:
Uy = 28.13 mm
If multiplied by the ductility coefficient, q= 1.5, these displacements become:

Uy = 28.13 x 1.5= 42.19 mm

They are now much smaller than the allowed displacements, [Ux] and [Uy] = 7.1 cm

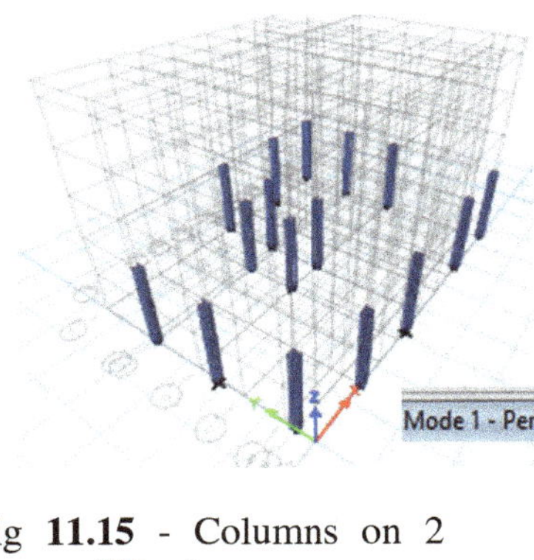

Fig **11.15** - Columns on 2 floors, 3D view

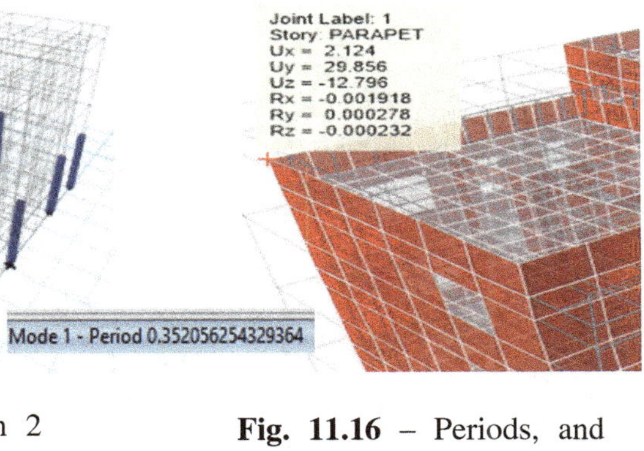

Fig. 11.16 – Periods, and displacements

b- The implementation of reinforcing columns has markedly diminished the compressive and tensile stresses across the masonry. Only a few isolated instances—specifically in the ELY combination—where stress values surpass the permissible limits exist; these occurrences are minimal compared to the scenario lacking reinforcement, highlighting the effectiveness of this structural enhancement.

Axis 2:

Max = 1.21 at [3.6, 4.4423, 0]; Min = -1.76 at [3.6, 0, 8.4]

Axis 4:

Max = 1.35 at [9.7, 13.87, 8.4]; Min = -1.61 at [9.7, 12.8672, 7.7778]

Axis 6:

Max = 1.51 at [18.42, 13.4188, 0]; Min = -1.86 at [18.42, 13.4188, 7.7778]

In conclusion, the results strongly support the efficacy of this solution, underscored by several compelling reasons:
a) it is economically viable, leveraging traditional construction materials like concrete and iron, which is particularly beneficial in countries with limited economic resources;
b) it enhances the first modal period while reducing displacements and surface stresses in the wall panels; and
c) the peripheral placement of the columns facilitates their implementation, promoting practicality in construction.

EPILOGUE

In concluding my multi-year analysis of building 77/5, I must acknowledge that it has been a significant challenge, and I consider my contribution modest. Given the importance of these buildings, my colleagues will continue to explore this area. This importance stems from the fact that

1- Many families have been housed within them and

2- Approximately 45 years have passed since the first of these types of structures were built, with the initial constructions taking place before the 1979 earthquake 6.5 Richter scale magnitude. There were known deficiencies in the technical conditions of buildings constructed before this earthquake.

3- Immediately following this earthquake, the government, through the Ministry of Construction and its associated institutes, began efforts to improve building codes, as reflected in the updated Technical Design Codes [KTP-N.2-89]

After many attempts to achieve a reliable result, I believe that the solution reached is a reasonable version of how this large contingent of buildings in our country can be reinforced most feasibly and economically possible. I hope that time will not disprove this conclusion that I have drawn! This was also the primary objective of my multi-year study and work.

My journey to this point began in 1979 with the defense of my thesis under the mentorship of Prof. Robert Progri, who was the first to inspire me to pursue further studies beyond my university graduation.

After 43 years of dedicated work, I can confidently assert that "learning is a lifelong journey, always accompanied by its adversary: forgetfulness!"

AUTHOR

REFERENCES:

[1] T.Paulay M.J.N. Priestley, Seismic Design of Concrete and Masonry Buildings
[2] B. Furiozzi, C. Messina, L. Paolini, Manual for Calculation of Structural Elements, New Edition, 2007
[3] Dr. Ing. Isai Clemente Structural Design Course, 2009/10, Lecturer at the University of Trieste [4] Niko Pojani, Seismic Engineering, Lecturer at the University of Tirana, 2003
[5] Msc Markel Baballeku, November 2014, Structural Damage Assessment in Education System Buildings, Tirana.
[6] Eurocodes EC6, EC8
[7] KTP-N.2-89. Technical Design Conditions, Academy of Sciences, Tirana 1989
[8] Magenes G., Kingsley G. R., Calvi G. M., 1995: " Static Testing of a Full-Scale, Two-Story Masonry Building: Test Procedure and Experimentally Measured Response". University of Pavia. [9] Calvi G. M., Kingsley G. R., Magenes G., 1996: " Testing of Masonry Structures for Seismic Assessment". Earthquake Spectrum, Vol. 12, No. 1
[10] Kerstin Lang, Shkurt 2002, Seismic Vulnerability of Existing Buildings, Zurich Institute of Structural Engineering, Swiss Federal Institute of Technology,
[11] Central Archive of the Ministry of Construction, Tirana
[12] Alberto Antonellia, Michele Bettia, Maria Luisa Del Saviob, Luciano Galanoa, Maurizio Orlandoa, 2010: " Methods for Seismic Analysis of Existing Masonry Buildings "
[13] Amir Hosein Salmanpour1*, Nebojsa Mojsilovic1 dhe Joseph Schwartz2 " Deformation capacity of unreinforced walls subjected to in-plane loading "
[14] R. Sabatino & G. Rizzano, Nonlinear Static Analysis of Masonry Structures with Simplified Equivalent Frames and Accurate Models, Department of Civil Engineering, University of Sales, Via Ponte Don Melillo, 84084 Fisciano (SA) - Itali
[15] ETABS® Version 18.1.1 2020, © Computers and Structures, Inc.
[16] Jan Červenka, Libor Jendele, Zdeněk Janda, Pragë, 12 Prill 2016 AMQUAKE, Cervenka Consulting Ltd., © 2010-2016
[17] Paolo Riva Department of Design and Technologies University of Bergamo February 2011 [18] High Strength Glass Fibers, World Headquarters/Americas, agy, Pub. No. LIT-2006-111R2 (02/06)

www.ingramcontent.com/pod-product-compliance
Lightning Source LLC
LaVergne TN
LVHW050302210926
848481LV00014BA/803